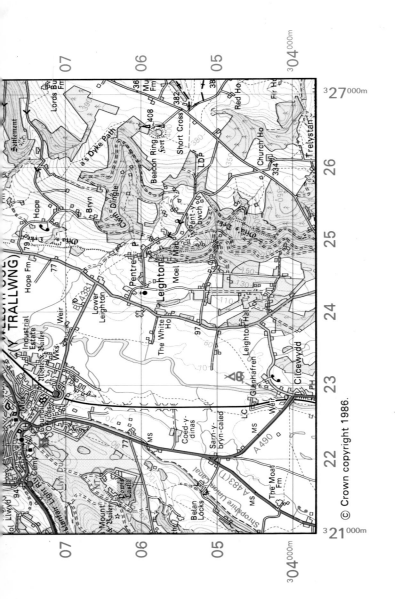

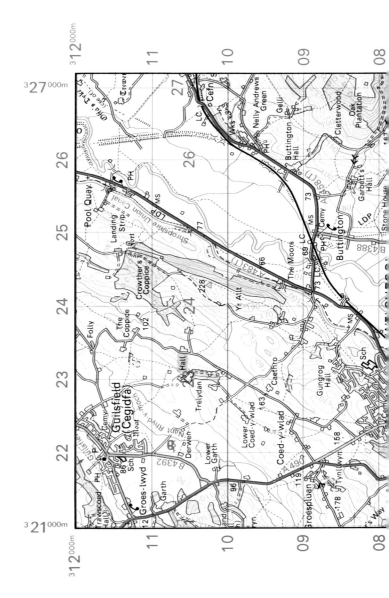

Make the Grade in GCSE Geography

David Jones

TEACH YOURSELF BOOKS
Hodder and Stoughton

First published 1987
Third impression 1988

Copyright © 1987
David Jones
Illustrations copyright © 1987
Hodder and Stoughton Ltd

No part of this publication may be reproduced
or transmitted in any form or by any means,
electronically or mechanically, including
photocopying, recording or any information
storage or retrieval system, without either
the prior permission in writing from the publisher
or a licence, permitting restricted copying,
issued by the Copyright Licensing Agency,
7 Ridgmount Street, London WC1E 7AA.

British Library Cataloguing in Publication Data
Jones, David P.
Make the grade in GCSE geography.–
(Teach yourself books).
1. Geography – Textbooks – 1945 –
I. Title
910   G128

ISBN 0 340 40127 3

Printed in Great Britain for
Hodder and Stoughton Educational,
a division of Hodder and Stoughton Ltd,
Mill Road, Dunton Green, Sevenoaks, Kent,
by Richard Clay Ltd., Bungay, Suffolk
Photoset by Rowland Phototypesetting Ltd
Bury St Edmunds, Suffolk

CONTENTS

Introduction

v

1 **Key Ideas and Skills**   1
1.1 Key ideas. 1.2 Atlases. 1.3 Map reading.
1.4 Graphs, diagrams and pictures.

2 **The Earth's Structure**   19
2.1 Continents and oceans. 2.2 Earthquakes and volcanoes. 2.3 Rocks and rock structures.

3 **Landforms and Landform Processes**   30
3.1 Rocks and landforms. 3.2 Weathering and slopes. 3.3 Rivers. 3.4 Ice. 3.5 Coasts.

4 **Weather and Climate**   51
4.1 Elements of Weather. 4.2 Temperature. 4.3 Rainfall. 4.4 Climate. 4.5 Special climates. 4.6 Weather systems. 4.7 Hydrological cycle and hydrology.

5 **Natural Environments**   65
5.1 Natural regions. 5.2 Hostile environments.

6 **People and Nature**   70
6.1 Hydrology, flooding and erosion. 6.2 Soil erosion and desertification. 6.3 Pollution. 6.4 Destruction of the natural environment. 6.5 Conservation and conflict.

7 **Population**   83
7.1 Population distribution. 7.2 Size, growth and structure of population. 7.3 Population movements.

8 **Settlement**   95
8.1 Types of settlement. 8.2 Site and situation. 8.3 Patterns and hierarchies of settlement. 8.4 Changes: decline and growth.

9 **Urbanisation**   105
9.1 Urban trends. 9.2 Patterns within cities. 9.3 People in cities. 9.4 Urban problems and planning. 9.5 Cites in less developed countries.

10 **Transport**   119
10.1 Distance, journeys and types of transport. 10.2 Routes and networks. 10.3 Transport planning and policies.

iv *Contents*

11	**Industry**	125
	11.1 Types of industry. 11.2 Factors of location. 11.3 Primary activities. 11.4 Secondary activities (manufacturing). 11.5 Tertiary activities (services). 11.6 Industrial regions.	
12	**Agriculture**	146
	12.1 Agricultural systems. 12.2 Factors affecting land use. 12.3 Patterns of land use. 12.4 Changes, problems and developments.	
13	**Resources**	157
	13.1 Types of resources. 13.2 Energy resources. 13.3 Conservation of landscapes and wildlife.	
14	**Contrasts in Development**	165
	14.1 World patterns. 14.2 Contrasts within countries.	
15	**Regions**	173
16	**Making the Grade**	176
	16.1 Projects. 16.2 Preparing for the examination. 16.3 Revision. 16.4 In the examination.	

INTRODUCTION

Syllabus
Geography syllabuses are all quite similar. Most are based on the same topics, although these may be grouped together in different ways or even given different titles. It is important that you find out what your own syllabus includes so that you can check that you have covered the work. Find out from your teacher what syllabus you are taking.

Topics
All syllabuses deal with the following topics: climate and weather, landforms, population, settlement, urbanisation, agriculture and industry. Topics such as resources and conservation, environmental problems, and contrasts between developed and developing countries are not always listed separately; however, you are still expected to have covered them within the topics listed. Most syllabuses are based on topics, but some specify the area of the world you have to study. In addition, your teacher may base your course on the study of a few specific regions.

Learning geography
The topics you study form the framework of the course. Within that framework, you need to know exactly what you have to be able to do. There are three 'sets' for you to master:

1 Facts   You have to know certain specific facts, particularly in the case of detailed examples you have studied. You will be expected to have studied examples at different scales, including local, regional, national, international and world scales. It is likely that there will be a slightly greater emphasis on examples from the British Isles compared with the rest of the world.

2 Ideas   You must understand the main ideas or principles to do with each of the topics in your syllabus. You will also be expected to show that you understand them by applying them to new situations.

3 Skills   These include being able to use an atlas, map reading, interpreting photographs, graphs and diagrams. You should be able to apply these skills and your understanding of ideas in a field-study project of your own as part of the assessment for the examination.

   A wide general knowledge will help you. Be alert to places in the news, as there is an element of geography in most news items.

Methods of assessment

All syllabuses will be examined or assessed in several ways. At the end of the course is the final examination. This will consist of one or two papers and these may contain different types of questions:

1 Short-answer or one-word-answer questions, or multiple-choice with perhaps 50 or 60 questions to be answered.
2 More detailed questions, where you have to answer three or four out of a selection of eight or more. You usually have 30 to 35 minutes in which to answer each question of this type.

During the course, you will be assessed on some aspect of course work. This assessment can take several forms, but will almost certainly include a fieldwork project of some sort. This will test your ability to organise and complete an enquiry into a geographical question within a small area. Some syllabuses also include coursework assessment based on class work. *Do not waste this chance to do well.* With course work, you have the opportunity to show initiative and to take extra time and care which will ensure good marks. If you throw away marks here, you will be at a disadvantage later on, when you come to tackle the written paper(s).

The proportion of marks given to the different elements of the examination varies. Course work generally accounts for 20 to 50 per cent and the final examination paper(s) the remaining 50 to 80 per cent.

Introduction vii

Plan of this book
The first few sections are general: they look at ideas and skills. Most of the book is taken up by sections on the main topics. The detail varies from one topic to another, but all the main points are covered and there are plenty of examples of the different types of questions. At the end of the book are sections on preparing for the final examination and the planning of projects.

Using the book
There is no set way to use this book. You should find it useful to read the introductory sections on ideas and skills early on in your course and to study them again several times. The last sections to do with examination questions and the written papers will be useful before internal school examinations, but most important in the last few months. The bulk of the book can be followed in many ways: use it throughout your course to supplement your school work and use the exercises to test yourself. The order in which you use these sections depends on the order in which you study the topics in school. For revision, study each topic in turn, but remember that because topics overlap, some work will appear under one heading in this book but under a different heading in your school work. Note also that many of the sample questions contain detailed examples and you could use the information given in answering other questions in your examination.

A number of questions and examples are based on the Ordnance Survey map extract reproduced inside the front cover of the book.

KEY IDEAS AND SKILLS
1.1 Key ideas

1

Showing information

Information about places can be shown on maps in different ways. The simplest are: Dots, Lines, Areas.

How features are shown depends on the **scale**. For example, in a study of a town the whole urban area would be shown as a patch or set of patches crossed by lines. However, in a study of the pattern of towns in a region of the country, the same town would be shown as a dot (Fig. 1.1). By using different colours or types of shading and different types of dots and lines, a vast amount of information can be presented on a map. By comparing maps from different times, changes can be seen (changes over time can also be shown on graphs, of course).

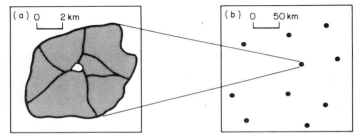

Fig. 1.1 *The effect of scale on the presentation of a town*

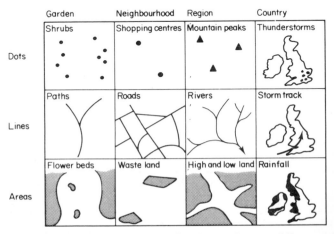

Fig. 1.2 *Examples of features shown as dots, lines and areas at different scales*

Key ideas 2

The idea of dots, lines and areas is important at the first stage of describing and explaining what places are like. If you think about it, anything can be shown using these. Fig. 1.2 shows some examples.

QUESTION 1.1 What dot, line and area features are shown below? Which area features are related to each other? Which area do the roads and villages avoid?

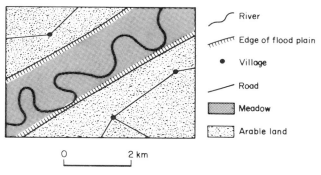

Fig. 1.3

The answers to the first question form part of a *description* and the answers to the second and third questions identify *relationships*. The next step is *explanation*. Can you suggest reasons for the relationships?

Key ideas

In geography, there are some very important ideas that apply in almost all branches of the subject. These form the framework for studying particular examples. The list of key ideas is potentially very large, but the following are the really important ones. You have certainly met them in your earlier work in geography, even if you did not use these actual words. Throughout this book, you will be using these ideas in connection with different topics and different examples.

Each of these can form the basis of a question about the geography of an area or the geography of a particular set of features. Here we shall look at a few applications of these ideas.

Location Where something is located is important. For example, a farm located at 200 metres is very different from one at 25 metres above sea level in the same region of Britain; a walker stranded at 1000 metres in the Cairngorms in winter is in a very different situation from someone in the middle of a park in London. You can use *latitude and longitude* or *grid references* to give a precise location. Otherwise, location can be described using distance and direction. How far is a place and in what direction?

3 Key ideas

Using words, there are many other points that can be made about locations. Here are a few: a village located on a river terrace, a lake in a narrow valley, a bridge at a narrow point on a river, a flood on lowland, and so on. Of course, all locations can be shown on maps.

Distribution This refers to a set or sets of features in an area. The area could be a school playground or the entire world. The features could be pupils at a point in time or the world's volcanoes. Whatever the features are, their pattern of distribution can be described. Fig. 1.4 shows some typical patterns of dot features.

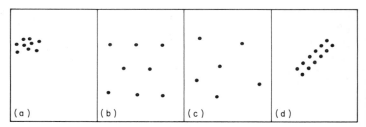

Fig. 1.4 *Dot patterns: (a) Clustered, e.g. people at a picnic site (b) Regular, e.g. farms on newly reclaimed land (c) Random, e.g. villages in a lowland area (d) Linear, e.g. buildings in a linear village*

Maps showing distributions can be compared. This enables us to identify *relationships* between patterns. For example, a comparison of maps of volcanoes and earthquakes shows a strong relationship: that is, a great similarity.

Interaction Examples of interaction are links between places, such as journeys to work or the movement of produce from farm to market. The idea involves movement between places and a system of links.

Distance This is the simplest and often the most overlooked idea in geography. There are different ways of measuring distance – in terms of time, cost or difficulty, for example, as well as linear distance measured in kilometres.

Hierarchy Many features are arranged or organised in a system called a hierarchy. This incorporates a large number of low-order or less important or smaller features; fewer middle-order, more important, bigger features; and very few high-order, important, large features. In geography the idea applies to places, but the idea is used more widely. For example, businesses are organised on the same basis.

Fig. 1.5 shows two examples of hierarchies, one from physical geography and one from human geography. Note how there are many low-order features (villages and small rivers), fewer middle-order features (town and larger rivers) and only one highest-order feature (city and major river).

Key ideas 4

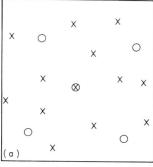
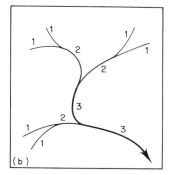

⊗ City
○ Town
× Village

Small rivers (1) join to form larger rivers (2) which join to make the major rivers (3) of the area

Fig. 1.5 *Hierarchies of (a) settlements and (b) rivers*

System Nothing is static. Movement and change mean that there are links. These links are arranged to make a system. A system has *inputs* and *outputs*. For example, the inputs of the hydrological cycle in a river basin are rainfall and sunshine, and the outputs are evaporation and river flow. Systems can be relatively simple (e.g. to show how an individual farm or factory works), or they can be complex (e.g. to show the working of a major city).

Region Areas of the earth's surface which have common features throughout and can be separated from surrounding areas on that basis are called regions. They can be defined in many ways. Some are physical regions which share the same relief and structural features, others are based on land use and share, for example, the same systems of agriculture; yet others are economic and share similar levels of economic development.

Scale This is a very basic idea. The scale of study affects all other aspects. A world wide study ignores local or even regional details, whereas a local scale study is concerned with very small scale differences which would be lost in a larger study.

KEY IDEAS AND SKILLS
1.2 Atlases

1

Using an atlas
An atlas is an essential reference tool for geographers. It provides basic information about the location and distribution of places and features and it often contains much other information in the form of statistics, case studies and explanations.

Some syllabuses allow candidates to use an atlas in the examination. This is no help unless you know how to make the most of the atlas:

1   You should be familiar with the layout of your atlas so that you can find particular maps quickly.
2   You must learn how to use the contents and index to most advantage.
3   You must learn to use the most appropriate map for the purpose.

Latitude and longitude
To find particular places, you need to understand latitude and longitude. Latitude gives the position of a place north or south of the equator in degrees. Longitude gives the position east or west of the prime meridian in degrees (Fig. 1.6).

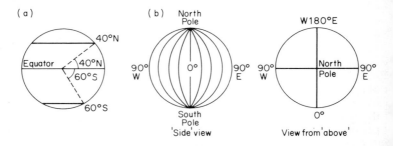

Fig. 1.6   *(a) Latitude (b) Longitude*

The method of locating a place is the same whether you use a map of the world, a map of a continent or a map of a small region. Note, however, that a small-scale map like a map of the world will only show large and important places and features.

The combination of latitude and longitude gives the co-ordinates of a place. Latitude is always given first and longitude second. Thus, the position of New Orleans is 30°N 90°W (Fig. 1.7). More precise locations are given by dividing degrees into minutes (60 minutes in one degree). The position of Moscow is therefore 55° 45′N 37° 35′E.

Atlases 6

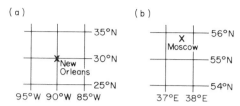

Fig. 1.7 *Locations of (a) New Orleans and (b) Moscow*

QUESTION 1.2 Using your own atlas, identify the places marked 1–5 on Fig. 1.8.

QUESTION 1.3 State the latitude and longitude of the places marked 6–10 on Fig. 1.8.

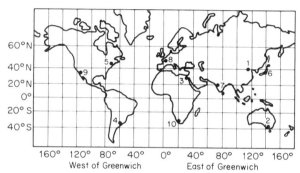

Fig. 1.8 *Latitude and longitude (Mercator map projection)*

Map projections

Your atlas might have a section explaining map projections. It is important that you understand a few basic points about the way maps of the world and parts of the world are drawn.

As the earth is spherical in shape, it is impossible for a map of a large area of the earth's surface to be without distortion. You cannot make a curved surface flat without stretching it or compressing it in some parts. Therefore, all maps are to be used with care. Some maps are drawn so that shapes are correct, in which case the areas are wrong. The *Mercator* map projection (Fig. 1.8) is a well-known one of this sort. It was devised to be used in navigation and should never be used to compare the areas of different parts of the world. Other projections are called *equal area*. They keep the correct relationship in area between different parts of the world. With these, however, the shapes are usually greatly distorted, especially towards the edge of the map. The *Mollweide* projection (Fig. 1.9(*a*)) is a commonly used example of this sort of map projection. Note that the lines of latitude and longitude are curved.

7 Atlases

Smaller areas of the world are often shown using map projections which are a compromise. The shapes are almost correct and the relative areas are almost correct. A common type is called a conical projection (Fig. 1.9(b)). The USSR is usually shown using this kind of projection, and because the country covers such an enormous extent in latitude the curvature of the lines of latitude is very marked. You will not be examined on map projections, but if you appreciate the differences you could avoid making simple errors.

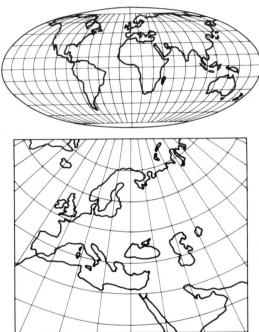

Fig. 1.9 *Map projections: (a) Mollweide (equal area) (b) Conical (see also Fig. 1.8)*

QUESTION 1.4 Study different maps in your atlas. Note the way particular areas of land differ in size and shape according to the map projection.

Longitude and time

Most atlases show the international time zones and a few explain the relationship between longitude and time. Local time varies according to longitude. Everywhere along a particular line of longitude has the same local time. For example, at midday the sun is at its highest everywhere along a line of longitude.

Atlases

The earth rotates from west to east, so places to the east are ahead in time and places to the west are behind. For example, if it is midday in London it is late afternoon in Moscow and early morning in New York. To put exact figures to it, the earth rotates once in 24 hours – that is, 360° in 24 hours or 15° in 1 hour and 1° in 4 minutes. For example, the longitude of Dacca is 90°E. If it is 11 a.m. in London (which is 0°), the time at Dacca will be 5 p.m. This is because there is a six-hour time difference (90° ÷ 15°) and being east of London, Dacca is ahead.

QUESTION 1.5 If it is 12 noon in London, what is the time at (a) 90°W, (b) 15°E, (c) 135°W, (d) 80°W, and (e) 120°E?

The answers to these questions give *local time*. It would be very inconvenient to use local time everywhere, however, as places a few kilometres apart would be operating on slightly different times. Therefore international **time zones** have been agreed, so that time within a longitudinal zone is the same, with changes (usually in one-hour steps) along agreed lines.

The *international date line* is part of this world-wide time system. If you start at 0° and work eastwards and westwards to 180°, you will find that there is a 24-hour or one-day 'jump' at 180°. Crossing from east to west you step back a day, and stepping from west to east you jump forward a day. This is on paper only, of course.

Sketch maps

Not many examination questions *ask* you to draw sketch maps, but there are good reasons for being able to do so:

1 A sketch map, properly labelled, is a quick way of showing what you know, saving a great deal of writing.
2 It is an excellent way to summarise information while revising.

A sketch map is not an accurate map. Do not try to show details. If a coastline is rugged, for example, draw a smooth curve and label it 'rugged coastline'. Proportions and general shape are what count. Remember to show an approximate scale if you can. Also note that if you are allowed an atlas in the examination, you can use it to help you draw quick, relevant sketch maps.

Fig. 1.10 *Sketch maps*

KEY IDEAS AND SKILLS
1.3 Map reading

Some examinations test map-reading skills directly in a map-reading question. Others include the use of maps as part of questions on a variety of topics. Whichever approach is used, the map-reading element is important and straightforward. The maps you have to use are the Ordnance Survey 1:50 000 (Landranger Series) and 1:25 000 (Pathfinder Series). Buy your own copy of your local map (either scale) and study it carefully. Although map-reading skills can be learned indoors, to develop them properly you should use your map for walking, cycling, bus, train and car trips.

What do you have to do in map reading?
1   Locate places precisely by (*a*) grid references (six- or four-figure), (*b*) direction and distance from a known spot, (*c*) marking them on an outline map, (*d*) marking them on a cross-section.
2   State direction using (*a*) the points of the compass and (*b*) degrees.
3   State distance by using the map scale in kilometres and metres.
4   Recognise map symbols – that is, name the features shown by symbols (which vary according to the map).
5   Recognise a whole variety of features, including physical and human features.
6   Work out the gradient of slopes.
7   Identify areas of differing height and slope.
8   Draw an accurate cross-section or a sketch cross-section or, more likely, use a cross-section.
9   Identify on a map features shown on a photograph.

These basic skills in map reading enable you to do much more. Using them, you can answer lots of questions about an area. The questions are usually based on the main topics you study in geography. You have to answer them using only the information on the map and your understanding of the topic.

Example   Describe the route followed by the road from Leighton Hall to Buttington.

The answer should include these points: the road follows the edge of the flood plain of the River Severn; it is generally at least 10 metres above the flood plain or valley floor, but keeps off the very steep slopes of the valley side.

A question which also asked you to *explain* the route would need an answer which referred to the fact that the route avoids land liable to flooding and steep land which would cause construction difficulties.

Make sure you always follow the instructions in a question (see p. 183 for the different words used in questions). Examples of map-reading questions are included in the rest of this section. When you answer them, make sure you give all the relevant information. For instance, on the map extract, Welshpool is clearly the largest settlement, but you can also say that it covers approximately 1.5 km².

Locating places using grid references

The grid reference system is based on the grid of lines shown on Ordnance Survey maps of all scales. On the maps we are concerned with, the grid lines are drawn 1 kilometre apart so each square covers 1 km². The lines running north to south are called *eastings* because they tell you how far east a place is. The lines that run east to west are called *northings* because they tell you how far north a place is (Fig. 1.11).

If you want to refer to a feature which occurs in a particular grid square you can give a *four-figure grid reference*. For example, in square 2510 of Fig. 1.11(a) is the confluence of a small stream with the River Severn. You identify the square by giving the figures for its south-west (or bottom-left) corner. Remember that a four-figure grid reference refers to a whole square. To give the location of a point, you use a *six-figure grid reference*. For this, each grid line is given as three figures, so easting 25 is referred to as easting 250. The space between eastings and northings are divided into ten so that 254105 in the precise location of the confluence (see Fig. 1.11(b)).

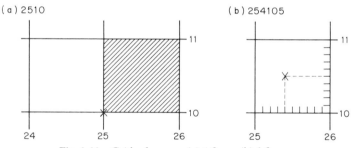

Fig. 1.11 *Grid references: (a) 4-figure (b) 6-figure*

At first it is useful to use a ruler to measure grid references. With a little practice, however, it is easy to estimate accurately. Remember that if you give the six-figure grid reference, be sure to give it for the feature itself and not the name that may be printed on the map. For example, Lower Leighton is at grid reference (or GR) 244065 on the Ordnance Survey map extract, because the reference refers to the buildings. The name is printed beside the place.

Locating places on an outline map Sometimes you will be given an outline map with only a few features marked, and asked to locate certain other features. This should be done precisely. Use the grid lines to work out the exact position, even if the outline map is drawn to a smaller scale.

11 Map reading

QUESTION 1.6 Mark on the map below the position of the bridge where the B4381 crosses the River Severn.

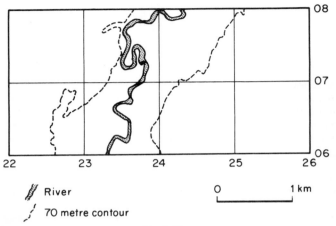

Fig. 1.12

2 Locating places by direction

The location of one place from another is frequently given by direction (or orientation), using the points of the compass. On the map extract, the railway from Welshpool, for example, runs north-east through Buttington. Welshpool is west north-west from Beacon Ring (264058). As well as using a compass rose, you can give a bearing. You do this with a protractor. Taking grid north as your north, you measure in degrees in a clockwise direction. Thus, Buttington is 45° from Welshpool station (Fig. 1.13(b)).

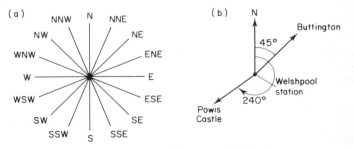

Fig. 1.13 *(a) Points of the compass (b) Bearings*

3 Measuring distances

The first thing you should note is the *scale* of the map you are using. This is

Map reading

shown in three ways: (a) in a statement, (b) as a scale line, and (c) as a representative fraction.

Fig. 1.14 *Methods of showing scale*

The representative fraction (or RF) is the only one to need explaining. In the example in Fig. 1.14, 1 cm on the map represents 50 000 cm on the ground. 50 000 cm is ½ km, therefore the scale is 2 cm to 1 km. The other scale commonly used is 1:25 000, that is, 1 cm representing ¼ km or 4 cm to 1 km. As you can see, a scale of 1:25 000 is a larger scale than 1:50 000.

QUESTION 1.7 What is the scale of a map with an RF of 1:100 000?

When asked to measure distances between places, you must give the answer in kilometres and metres. The direct distance between points can be measured with a ruler in centimetres and divided by two or four depending on the scale of the map being used. Distance along a road should be measured using dividers set to a small distance. Convert the total distance measured to kilometres, using the scale of the map.

QUESTION 1.8 On the Ordnance Survey map extract, what is the distance from Welshpool station to the road junction in Buttington (*a*) in a straight line, and (*b*) along the B4381 and B4388?

Recognising map symbols

The symbols used to show particular features are called map symbols or conventional signs. The symbols vary for different scale maps. You must learn them carefully. Many symbols are quite obvious in their meaning but there are some particularly awkward ones, especially boundary lines of various sorts and footpaths. The best way to learn symbols is by *using* your own map.

QUESTION 1.9 On the map extract, what do the symbols at the following locations mean?

(*a*) 230044, (*b*) 255041, (*c*) 251097, (*d*) 221065, (*e*) 247099.

Recognising geographical features

This links up with all the work you do in geography. Almost all the topics you study should include examples at 1:50 000 and 1:25 000 scales. Map reading tests your knowledge of the features and your ability to recognise them. For convenience, they are divided into physical features and human features. As you study each topic, you should learn how the features appear on maps.

13 *Map reading*

Physical features include:

   landforms produced by rivers, e.g. flood plains, waterfalls
   landforms produced by ice, e.g. corries, glacial troughs
   coastal landforms, e.g. cliffs, spits
   landforms related to geology, e.g. scarp slopes
   uplands and lowlands
   drainage systems
   'natural' vegetation, e.g. rough grassland, heath

Human features include:

   settlements
   transport, e.g. roads, railways
   industry, e.g. works, quarry
   agriculture and rural land use, e.g. orchards, farms
   tourism

Exercises tend to ask either for specific examples (in which case you mu⸺ give grid references and be precise), or for a description of certain feature⸺ Examples of the second type are included in later sections of the book.

QUESTION 1.10 (*a*) Using the map extract, state the grid reference of ⸺ example of each of the following: level crossing, river confluence, riv⸺ meander, a hamlet. (*b*) Describe the land use in the area immediately sout⸺ west of Welshpool.

6 Gradient

The slope of the land – or gradient – is sometimes asked for, and is simple⸺ work out. For example, the spot height east of Buttington at 255089 is at ⸺ metres. The height of the road at Cletterward at 262085 is 160 metres. Th⸺ gives a height difference of 87 metres. The horizontal distance between t⸺ two points is 800 metres.

$$\text{gradient} = \frac{\text{difference in height}}{\text{horizontal distance}}$$

so $^{87}/_{800} = ^{1}/_{9 \cdot 2}$ or approximately 1 in 9

On new road signs, gradient is given as a percentage and it may be giv⸺ like that in an examination. As with any other calculation, the change ⸺ simple:

$$\text{gradient} = \frac{\text{difference in height}}{\text{horizontal distance}} \times 100$$

so $^{87}/_{800} \times 100 = 10.9\%$ or 11% approximately

Note the map symbols for steep road gradients of 1 in 5 and steeper, a⸺ gradients between 1 in 5 and 1 in 7.

Map reading 14

QUESTION **1.11** On the map extract, compare the gradient between 250080 and 255075 with that between 255075 and 260070.

7 Identifying areas of different height and slope: relief

This goes with the last two sections. You need to use the contours, spot heights and triangulation points to identify shapes and heights, and you need to use your knowledge of different physical features. Remember that contours are lines joining places of the same height, spot heights are dots with the height written alongside and triangulation points are shown by triangles with the height alongside.

There are two stages in describing relief:

1 Separate areas which are clearly different, as shown by the heights and the pattern of contours. On the map extract, the Severn Valley is wide and flat as the absence of contours shows. To the west is higher ground which is hilly (and broken up into smaller areas of high ground). To the east, the land forms one high mountain area – the Long Mountain. Its steeper slops are shown by the closeness of the contours. On the part of the summit shown on the extract around Beacon Ring, the contours are more widely spaced, generally indicating more level slopes.

2 Look in more detail at the steepness of the slopes and their shapes. On the west side of the Severn Valley, the contours are more widely spaced than on the slopes of the Long Mountain – showing less steep slopes. On the Long Mountain itself, the shape of the slope rising from Lower Leighton (244065) is gentle at first and then becomes steeper before gradually levelling off at the top.

QUESTION **1.12** Imagine you were able to walk from west to east along northing 070. Describe the slopes, noting whether you would be walking up or down hill, whether the slopes are gentle or steep, and how they change.

8 Drawing and using cross-sections

You are not often asked to draw a section, but it is useful to be able to draw a sketch section in an examination and you may need to draw an accurate section as part of a project or other piece of coursework.

To draw an accurate section Mark the ends of the section on the map, then fold a piece of paper and lay it along the line joining the two points. Mark the ends of the section on the paper, mark each contour that crosses the paper and number the contours as you work along the line. If the slopes are very steep, it may be easier to mark only the dark/heavy contours. You will almost certainly be given the framework for drawing the section. Lay your paper along the base line and transfer your points to the correct height on the framework. Join your points to complete the section.

To draw a sketch section Draw a framework with approximate heights marked, then draw the section freehand. It helps to mark out key points,

15 Map reading

such as the tops and bottoms of slopes, in order to keep the horizontal scale as accurate as possible.

QUESTION 1.13 Fig. 1.15 shows an accurate section. Using the Ordnance Survey extract, draw a sketch section on the blank framework. Note that the grid references for the ends of the sections are marked on the framework.

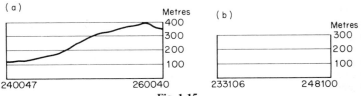

Fig. 1.15

Using a cross-section is more likely to be part of a question. It usually involves adding labelling to a section. This is very straightforward providing you are accurate and use the scale carefully. If the scale is the same as your map, you can measure directly from the map to the section. If the scale is different, remember to adjust your measurements. Labels should be written above with arrows dropping down to the correct point.

QUESTION 1.14 On the cross-section below, label the following: the A483(T), the B4388, Gunrog Hall, Cletterwood. (The River Severn is already shown.)

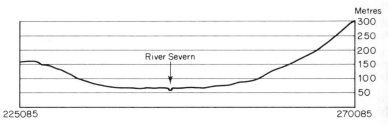

Fig. 1.16

9 Using a photograph

If you are given a photograph with a map-reading question, your first task is to identify the scene in the photograph on the map. Sometimes you are given the location from which the photograph was taken. In that case it is fairly easy to pick out some key features. This enables you to state the orientation of the photograph (the direction in which the camera was pointing). Note that the photographs used are usually either aerial or taken from high ground.

You may be asked to give the grid reference of a feature in the photograph. This requires careful examination of the photograph. This sort

Map reading 16

of question is testing the basic ability to recognise places on the map and on the ground or photograph.

Other questions depend on the photograph. Photographs showing landforms, land use, village sites or urban development may lead to more detailed questions to do with those topics. These may involve describing and explaining the scene using information in the photograph and on the map, together with your own knowledge of that topic.

For practice, make frequent use of photographs with maps or go out with your own map and study an area from a high viewpoint.

Map-reading questions

The basic skills in map reading are tested by the straightforward questions on techniques. Parts of questions (and whole questions in some papers) use the map as a source of information on various topics. For example, questions could include:

QUESTION **1.15** Study the town of Welshpool on the Ordnance Survey map extract.
(a) Find, name and give the grid reference of the earliest evidence of settlement at Welshpool.
(b) The town centre – or business district – generally extends along the A483(T) through the town. What map evidence supports this statement?
(c) Explain why most of the town lies to the west of the railway line.

QUESTION **1.16** The settlements of Welshpool, Guilsfield and Buttington differ quiet considerably. Using the headings *size*, *form* and *functions*, show how the settlements differ.

QUESTION **1.17** Look carefully at Cwm Dingle (squares 2506 and 2606). Draw a labelled sketch section to show the shape of the valley at 260060.

QUESTION **1.18** Traffic problems are common in Welshpool during the summer months, especially at weekends.
(a) Where are the traffic problems likely to be centred? Give your answer in words *and* as a grid reference.
(b) Explain why traffic problems occur where you stated in (a).
(c) How is the town likely to be affected by these traffic problems?

As you can see, this type of question draws on your knowledge and understanding of all aspects of geography and tests your ability to apply them to new situations, as well as testing your basic map-reading skills.

Note the following points:

1 Unless extra information is provided, use information given on the map only.
2 Suggest possible explanations rather than stating definite causes.
3 Give full descriptions and use the correct terms for features.
4 Look for relationships (e.g. between low, flat land and flooding risk; or steep, high land and lack of settlement; or woodland found only on steep slopes).

KEY IDEAS AND SKILLS
1.4 Graphs, diagrams and pictures

These serve two purposes. In examination questions, they provide information and ideas in a visual form, and they also provide a way of presenting your own information and ideas.

Graphs
Line graphs show how things change over time.

Proportional circles compare amounts or quantities in different places or at different times. When circles are divided (pie charts), they show the relative importance of different elements.

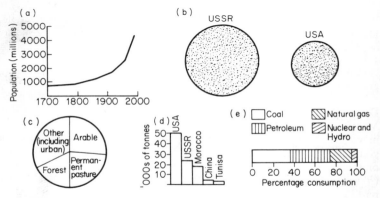

Fig. 1.17 *Types of graph: (a) Line graph (World population growth) (b) Proportional circles (Land areas of the USSR and USA) (c) Pie chart (Land use in Belgium) (d) Bar graph (Main producers of rock phosphate) (e) Divided column (Energy consumption in the UK)*

Bar graphs – the simplest form of all – can be used for all these purposes.

Scatter graphs are used to show cases where there is a relationship between the variables. The examples show a *positive* relationship and a *negative* or *inverse* relationship. Where most of the points on the graph are in a rough line, the relationship is clear.

The *description* of a graph is a straightforward task, but it does need care and attention to detail. Look at each of the graphs in Fig. 1.17 and see how many points you can make about each one.

The title of a graph, or its description in a question tells you what you have to describe. For example, the bar graph in Fig. 1.17 shows the main producing countries of rock phosphate. Which country is the main producer? Where else is rock phosphate produced in large quantities? (Note

Graphs, diagrams and pictures 18

whether the quantities are shown in number or percentage.) Make the same points for the other graphs. Remember to refer to the scale and to be precise. In an examination question the number of marks will be given. This is a guide to how many points you have to make.

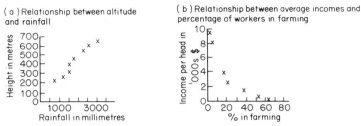

Fig. 1.18 *Scattergraphs showing (a) positive correlation (b) inverse or negative correlation*

In projects and individual studies, you must select the most appropriate graphs to illustrate and analyse the information you collect.

Diagrams, pictures and cartoons

Many ideas or sets of processes can be shown more clearly by diagrams than in words. It is often useful to incorporate diagrams in your own answers, if you can. More usually, you will be given diagrams and asked to comment on them. You may be asked to: (*a*) extract specific pieces of information, (*b*) work out how one part affects another, and (*c*) show how change in one part influences another.

Photographs

These are given not only as part of a map-reading question. The first task with a photograph question is observation. You must look at the photograph in a systematic way. What are you looking at? Does the caption (if there is one) give you a clue, or do you have to work it out for yourself? What is the main feature? Is it a human or physical feature? What aspect of geography is it about? If it is a physical feature, what particular feature or features are shown? What processes formed or are forming them? Are there any obvious relationships? If the photograph is to do with human geography, what topic is it concerned with? Is it industry, rural settlement, urban geography, transport or agriculture? In all cases, you have to *observe*, *record* your observations, *describe* the scene, *identify relationships between elements* of the scene, and *explain* the scene, possibly by referring to the **processes** at work.

THE EARTH'S STRUCTURE
2.1 Continents and oceans

The earth is one self-contained unit. Its surface is broken into land (30%) and ocean (70%). These are distributed unevenly, and most of the land masses are in the northern hemisphere.

You are expected to know the facts of the distribution of major land and sea areas as well as many of the smaller land and sea areas. Study an atlas so that you know where the oceans are deep or shallow, and where there are enclosed seas or large bays. With the land masses, you must recognise shapes, know where there are narrow necks of land joining land masses and so on.

Mountains, plateaus and plains
The continents vary considerably. On the large scale, mountains, plateaus and plains are the three sets of relief features. You should be familiar with their distribution on each of the continents. In South America, for example, we can see a range of mountains in the Andes, plateaus in the Brazilian Plateau and plains in the Pampas.

Plate tectonics
Hidden from view is another division of the earth. The earth's crust is divided into a number of rigid plates which are able to move about probably through being shifted by convection currents in the layers beneath the crust. The **plate boundaries** are marked by zones of volcanic and earthquake activity. There are three different kinds of plate boundary (see Fig. 2.1):

1 Constructive plate boundaries Here, two plates move apart, allowing new rock to well up from below the crust. New crust is being formed all the time and volcanic eruptions build up huge mountain ranges under the sea. In the Atlantic Ocean they form the Mid-Atlantic Ridge, parts of which break the surface as islands e.g. Iceland.

2 Destructive plate boundaries Here, two plates move towards each other, one being forced down below the other (such as in Japan). The part of the crust that is forced down (or *subducted*) is melted, and the molten material rises up through the crust. Some of it reaches the surface as volcanoes, but much of it solidifies a few kilometres below the surface. These form features called **batholiths**, which may eventually be exposed at the surface when the rock layers above have been eroded. Volcanoes are very common, and the 'Pacific Ring of Fire' illustrates this point. There are hundreds of volcanoes in the mountains along the west coast of South Central and North America and in the chains of islands off the east coast of Asia as well as in New Zealand. These volcanoes form an almost continuous ring around the Pacific Ocean.

Continents and oceans 20

3 Transform faults These occur where two plates slide against each other in opposite directions. The San Andreas Fault in California, where the Pacific plate is moving north against the North American plate, is an example of a transform fault.

1 Constructive plate boundary (section)

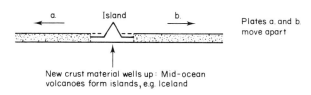

Plates a. and b. move apart

New crust material wells up: Mid-ocean volcanoes form islands, e.g. Iceland

2 Destructive plate boundary (section)

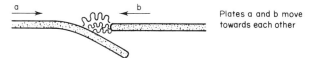

Plates a and b move towards each other

Plate a sinks beneath b. Sediments between the two plates are crumpled to form fold mountains, e.g. Andes. Plate b rock is heated at depth, becomes molten and lava rises to form volcanoes

3 Transform fault (map)

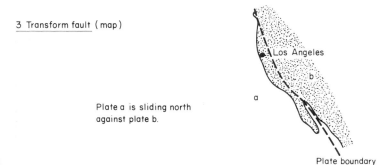

Plate a is sliding north against plate b.

Fig. 2.1 *Types of plate boundary*

The movement of plates is important because the boundaries mark the zones where earthquakes and volcanoes occur. Earthquakes are the result of rapid movements or vibrations caused by shifts along plate boundaries or faults, or the movement of volcanic lava beneath the surface.

21 Continents and oceans

Questions on this topic are not common and would probably be linked with earthquakes and volcanoes. Check your syllabus to see if plate tectonics are included.

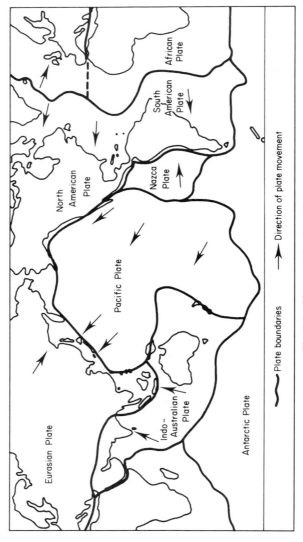

Fig. 2.2 *Plates and plate boundaries*

2

THE EARTH'S STRUCTURE
2.2 Earthquakes and volcanoes

Note the remarkable similarity of the patterns of volcanic and earthquake activity (Fig. 2.3) and the pattern of plate boundaries shown in Fig. 2.2.

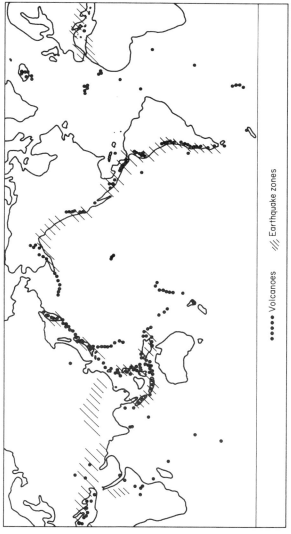

••••• Volcanoes   /// Earthquake zones

Fig. 2.3 *Earthquake zones and volcanoes*

23 Earthquakes and volcanoes

Both types of activity and their distribution have already been partially explained.

Volcanoes

A volcano is a vent or fissure in the earth's surface through which hot material from lower levels reaches the surface. Volcanoes can be *active*, *dormant* or *extinct*. An active volcano is one where eruptions may actually be occurring, or – more likely – where there is mild activity in the form of gases being emitted. A dormant volcano is one which has not erupted nor shown any sign of activity in historic time. An extinct volcano is one where there has been no activity for at least 25 000 years. Any volcano which has erupted in the last 25 000 years is regarded as potentially active.

When a volcano erupts, gas, lava and ash are thrown out in varying proportions, depending on the type of eruption. Some eruptions are mild and a fluid lava pours out as a 'river of fire'. Others throw out much ash and little else. The classic conical form of volcano occurs when ash and lava are put out alternately. The most dramatic eruptions are marked by violently

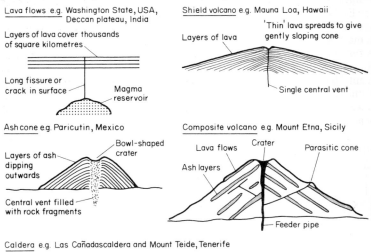

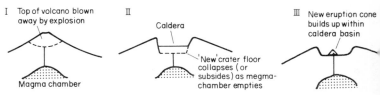

Fig. 2.4 *Volcanic landforms*

destructive explosions, as with Mount St Helen in the USA in 1980, and Krakatoa in Indonesia in 1883.

The hazard to human beings is fairly obvious: lava flows can destroy farmland and settlements and areas can be devasted by explosions. However, ash and gas are just as great a hazard. The Roman town of Pompeii was buried in ash and its citizens overcome by gas in AD 79. More recently, on the Icelandic island of Heimaey, the town was almost covered by ash and its inhabitants had to be evacuated. In addition, a lava flow almost blocked off the fishing harbour.

Volcanic landscapes

These are incredibly varied. Individual volcanoes have different shapes according to their origin, and in any one volcanic area you find a combination of individual landforms making up the present-day landscape (Fig. 2.4).

Tenerife, Canary Islands The bulk of the island is actually one volcanic mountain, Mount Teide, a composite volcanic cone. The peak and present cone lies within a large crater or *caldera* called Las Cañadas. The island has many lava flows of different ages, together with ash deposits produced by the main volcano and subsidiary vents. Many small ash cones occur, and are the result of eruptions from minor vents. Recent lava flows (within the last 500 years) have extended the land into the sea in places. The soil is exceptionally fertile as the weathering of the lava releases minerals. The porous nature of the ash allows rainwater to percolate underground, from where it is tapped for irrigation. The volcanic peak itself is a tourist attraction.

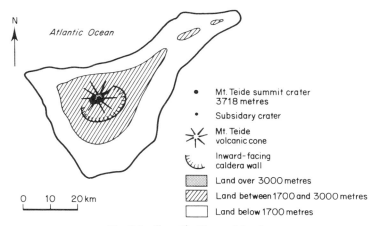

Fig. 2.5 *Tenerife, Canary Islands*

25   *Earthquakes and volcanoes*

Earthquakes

The actual movement causing an earthquake may be 10 to 30 kilometres below the surface, and is known as the **focus**. The location on the surface immediately above, which receives the greatest shock, is the **epicentre**. Shock waves travel outwards from the epicentre, and the damage caused by vibration decreases away from the epicentre. The strength of an earthquake is generally stated by its magnitude on the **Richter scale**. This is a measure of the power produced by the earthquake. Earthquakes are also compared according to the amount of damage done. Clearly, the greater the magnitude of the earthquake, the more widespread its effects.

Earthquakes have effects on surface features and landforms. They may: cause the rise or fall of areas of land by appreciable amounts; create a fault scarp; cause offsetting of streams and other features; cause landslides, especially in unstable mountainous areas – and where these dam a valley to create a lake, another potential hazard is created when the lake overflows. In the oceans, the rise and fall of the ocean floor sets in motion a series of waves called *tsunamis*, which can devastate coastal regions thousands of kilometres away when they reach shore.

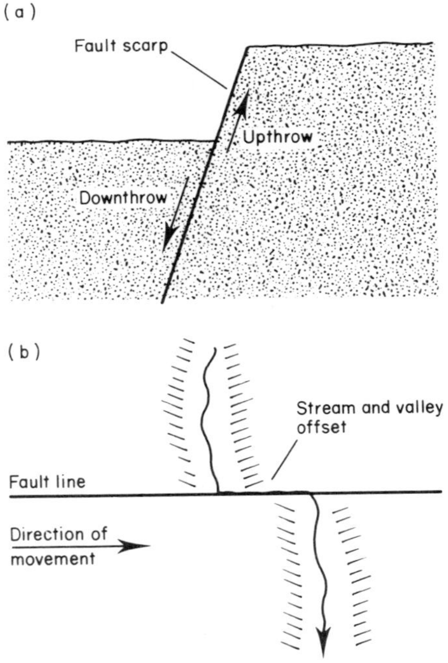

Fig. 2.6   *(a) Fault scarp (b) Offset stream and valley*

Earthquakes and volcanoes

The damage caused by an earthquake depends on the type of buildings in a populated area. Modern buildings can be made to withstand considerable earthquake shocks, but some traditional building materials, like sun-dried mud bricks, can collapse quickly and have caused tens of thousands of deaths in earthquakes.

The response to an earthquake (or any other disaster) is important, too. In less developed and isolated areas of the world, more people are likely to die, not only because of the use of traditional building materials, but because of the time taken for help to reach the area and the delay in treating the injured. The availability of highly-trained medical and rescue personnel is a key factor in saving the injured and dealing with the problems of survivors who are likely to be homeless.

QUESTION 2.1 Describe the pattern of earthquake and volcano zones. Explain why they are so similar.

QUESTION 2.2 Name an area of the world where there is volcanic activity.
(a) Describe the volcanic landforms found there.
(b) What effect has volcanic activity had on human activity?

QUESTION 2.3 Name one natural disaster.
(a) What were the immediate effects?
(b) What were the long-term problems?

THE EARTH'S STRUCTURE
2.3 Rocks and rock structures

The rocks of the earth's crust are important geographically because:

1 They have a major influence on the relief features of the surface.
2 They have economic value, as they contain minerals and fuels as well as the raw materials for building.

Rocks are classified into three main groups according to their origin:

1 **Igneous rocks** are formed from molten material from the lower levels of the earth's crust. They are *extrusive* or *volcanic* rocks (rocks resulting from volcanic eruptions), and consist of lava, ash and dust deposits. *Intrusive* rocks are formed beneath the surface, as molten material or *magma* did not reach the surface. The intrusive rocks either replace other rocks, fill spaces between them, or cut straight across them.

2 **Sedimentary rocks** are largely formed from layers of sediment deposited by rivers in seas and lakes. These sediments accumulate and are converted into rock as water is squeezed out, and as spaces are filled by calcium, iron or silica cement. The type of rock depends on the coarseness of the sediments. Thus, a rock made of coarse sand is a *sandstone* and one with fine layers of thin mud is a *shale*. Some sedimentary rocks are formed from the remains of living things, two of the most important being limestone and coal.

A quarry or cliff face shows the basic structures of sedimentary rocks. Each layer of rock or *bed* is separated from the next by a *bedding plane*. The beds themselves usually have vertical cracks called *joints*. The *strata* (the series of beds of rock) may consist of only one type of rock, or they may consist of alternating layers (e.g. sandstone and shale), or they may consist of a whole series of different types of sedimentary rocks.

3 **Metamorphic rocks** are formed from other rocks which have been changed in some way, usually by heat or pressure. Sometimes, they can be local in occurrence, such as rocks next to an igneous intrusion which have been baked by the heat. Where they cover large areas, they tend to be old rocks which have been subjected to heat and pressure many times through geological history.

Fig. 2.7 *Basic rock structures*

Rocks and rock structures

For a detailed study of rocks, you need to refer to a basic geology book. From your point of interest, however, there are two main features of rocks which are particularly significant geographically:

1 *Properties of the rocks* *Resistant* rocks (*not* hard) are less easily eroded than less resistant ones. *Permeable* rocks allow water to pass through, either along joint planes and bedding planes, or through pore spaces between individual particles. The latter are *porous* rocks. Rocks through which water cannot pass are *impermeable*.

These qualities are very important as they affect surface water and water storage. The effects of different rocks types on the scenery are studied in Section 3.1.

2 *Rock structure* The structure of the rocks refers to the way they are arranged and mostly concerns sedimentary rocks. Fig. 2.8 shows the different arrangements of rocks. Folds can be simple as in the Weald District of south-east England, or very complex as in the Alps. Faults, which are breaks in the rock layers are, like folds, brought about by movements of the earth's crust.

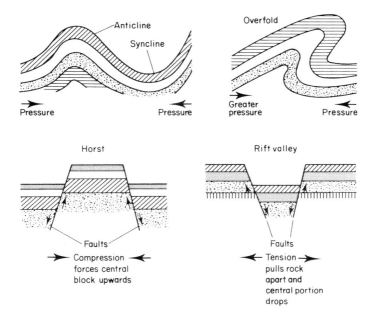

Fig. 2.8 *Rock structures*

Rocks and rock structures

The effects of rock structure on scenery and landforms are covered in the next section, but some important points to note are:

1 A fault creates a zone of weakness where shattered rocks can be eroded more easily.
2 A fault which brings resistant rock against less resistant rock creates conditions for more rapid erosion of the weaker rock, leaving the more resistant rock as higher land.
3 Folding of rocks results in anticlines being eroded more easily, as the rock is stretched by the folding.
4 Even small features like joints in the rock create local points of weakness.

Questions which use these ideas are generally parts of other questions. Ideas about permeability of rocks come into questions on hydrology (4.7 and 6.1) and water supply. Ideas about the resistance of rocks come into questions on landforms generally (3.1 to 3.5). Short questions or multiple-choice questions may test your knowledge of definitions of terms.

3 LANDFORMS AND LANDFORM PROCESSES
3.1 Rocks and landforms

In this topic you have to be able to name and describe the different sorts of landform features, and to describe and explain the processes which produce the landforms.

Recognition of general landform features, like the hills and lowlands, mountain ranges and river valleys of particular areas, is really a question of atlas work and general knowledge. However, you should have a good idea of where to find particular kinds of landforms and, for the British Isles at least, be able to name the main physical features.

Rocks and scenery
The type of rock and its structure affects scenery in general and individual landforms. Different categories of rocks and their characteristics were listed in Section 2.3, and you should learn them. In this section, we are going to look at the landscapes produced by a few kinds of rock type and a few kinds of rock structure.

Granite moors and sandstone moors
These are dealt with together because the scenery is often similar. Granite is a resistant igneous rock which is broken up into large blocks by joints. Many sandstones (and gritstones) are also resistant, and are also broken into large blocks by joints as well as widely-spaced bedding planes. Apart from the percolation of some water along the joint planes, the rocks are impervious and the generally level high surfaces form moorland areas cut by deeply incised valleys. *Tors* are massive blocks of bare rocks on exposed hills in granite areas, like Dartmoor, but they are also common on the sandstone and gritstone moors of the Peak District. The generally impervious character of the rock and resulting poor drainage also means that large areas of peat are common.

Limestone scenery
Limestone and its softer partner chalk have quite distinct features. There are several types of limestone, but they are all permeable so that there is a *lack of surface drainage*. Since most water goes underground, there is less surface erosion. This, together with the fact that limestones are generally fairly resistant and often very resistant, means that limestone areas almost always form uplands.

In Britain, Carboniferous Limestone forms extensive upland areas and produces some spectacular scenic features, both above and below ground (Fig. 3.1). It is a 'massive' rock, with widely spaced bedding planes and prominent joints. Rainwater is the agent by which the scenery is formed. As it passes through the air rain absorbs carbon dioxide, making a weak acid. When it percolates along the joint planes and bedding planes it dissolves the

31 Rocks and landforms

limestone, which is almost pure calcium carbonate. The joints are enlarged and, as a result, all water falling on the surface is rapidly channelled underground. The place where a stream goes underground is called a *sink* or *pothole*. *Caves* or *caverns* are formed by solution and erosion by the streams running through them. They have their own features: calcium carbonate is redeposited from water coming into them to form *stalactites* (hanging from the roof), *stalagmites* (on the cave floor), *columns* (where the two join) and *flowstone* (on the walls). On the surface, the landscape can be striking, with large amounts of white or greyish-white rock exposed at the surface. This is often called *limestone pavement*, with the enlarged joints (*grikes*) separating blocks of limestone (*clints*). Only the very deeply-cut valleys of large rivers have permanent water, and it is common to find narrow gorges which are dry and have dry waterfalls. As well as dry valleys cut by streams that have since gone underground, we find deep gorges which are the result of the collapse of cave systems. Limestone scenery is also known as *karst* scenery, after the limestone area of Yugoslavia.

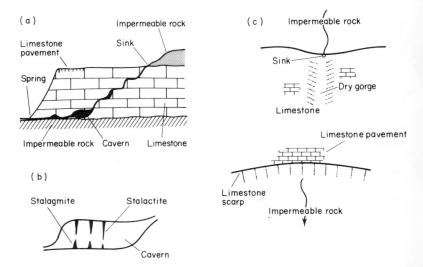

Fig. 3.1 *Limestone scenery: (a) Cross-section (b) Enlargement of cavern (c) Map*

Chalk scenery

Chalk is a relatively soft form of limestone. It forms uplands, but there are no chalk hills higher than about 300 metres. Dry valleys are widespread and the land generally has a rolling form. On Ordnance Survey maps, the appearance of chalk areas is quite distinctive. There are no sinks, as in Carboniferous Limestone, because the rock is porous and water percolates through the body of the rock very quickly. The water soaks down into the rock and the lower levels tend to be saturated. The top of this saturated layer is called the *water table*. If a valley has been cut down deeply enough to reach the water table, a permanent stream will flow – but this only applies to the lower reaches of most valleys. Changes in water levels with wetter seasons sometimes results in streams flowing where the valley is normally dry (hence the name Winterbourne – 'bourne' meaning stream).

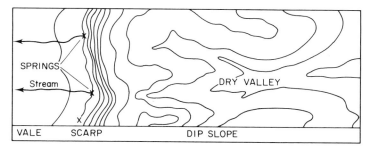

Fig. 3.2 *Contour map of Chalk scenery*

The other characteristic features of chalk scenery in Britain are clearly linked to the structure of the rocks. The chalk and other adjacent rocks are usually dipping, and the alternation of more resistant rock (chalk) with less resistant rock (clay) gives rise to *scarp and vale* topography, or an alternating series of hills and lowlands. The hills generally have a steep or *scarp* slope overlooking the vale and a more gentle *dip slope* on the other side leading down to the next vale (Fig. 3.3).

This pattern gives rise to a line of springs at the foot of the chalk scarp, where the impermeable clay prevents the water percolating further downwards.

Rocks and underground water

Chalk is an excellent example of a rock which stores water. Such rocks are called *aquifers*. The structure of the Thames Basin shows a feature called an *artesian basin*, in which a water-bearing rock is trapped between two

33 Rocks and landforms

impermeable layers. This water can be tapped by boreholes drilled down through the capping layer. In this case, the rocks are arranged in a downfold or *syncline*. This is one of the most common rock structures to form an artesian basin. Fig. 3.4 shows the Thames Basin syncline and the relationship of scarp and dip slopes to the overall arrangement of the rocks.

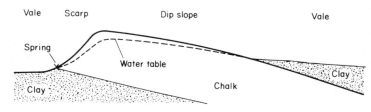

Fig. 3.3 *Scarp and vale topography*

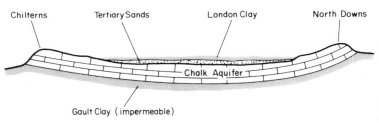

Fig. 3.4 *The Thames Basin*

QUESTION 3.1 Explain why there is a line of springs at the foot of the scarp on the map in Fig. 3.2.

QUESTION 3.2 Study Fig. 3.4.
(a) What geological structure is shown?
(b) On the section, mark and label a scarp slope, a dip slope and a water table.

QUESTION 3.3 Study Fig. 3.1.
(a) Explain how and why the stream goes underground at the point marked 'sink and reappears at the point marked 'spring'.
(b) What evidence is there to show that the stream once flowed across the surface of the limestone?
(c) Make a labelled drawing to show the features of limestone pavement and to explain how it is formed.

3   LANDFORMS AND LANDFORM PROCESSES
3.2 Weathering and slopes

Rocks are broken down by *weathering*, either producing rock debris or dissolving the rock in solution.

Physical or mechanical weathering breaks up exposed rock, principally by freeze-thaw action or frost shattering. Water in the cracks on the surface of the rock freezes and expands as ice forms. This exerts pressure, forcing the crack open. Eventually, repeated freezing and thawing causes pieces of rock to break off. Rock debris loosened from a rock face in this way falls and gradually accumulates to form a scree slope (Fig. 3.5(*a*)). In a similar way, the roots of plants force open cracks in rock.

Chemical weathering occurs when rain water, which is a weak acid, or acids in the soil attack the rock and cause it to break down. In limestones, which are largely composed of calcium carbonate, there is only a small residue left from weathering to form soil, because most is removed in solution. In other rocks, chemical weathering breaks up the cement holding the particles together and thereby aids the development of a deeper soil. Chemical weathering is active in cities, where decaying stonework in old buildings is the direct result of pollution of the atmosphere by chemicals.

The relative importance of the two types of weathering varies. In general, it is true to say that where there is bare rock, physical weathering is more important; where there is a soil cover, chemical weathering is more active.

The movement of weather material is called *mass wasting*. This covers all forms of movement of soil and rock material down slopes. Even on gentle slopes, soil particles are moved downhill by a process called *soil creep* (in which particles of soil are displaced downhill by rain). Even though it is a slow process, its long-term effects can be marked (Fig. 3.5(*b*)).

When soil becomes waterlogged after periods of heavy rain, *soil flows*, *soil slides* and *mud flows* can occur. These are the result of the structure of the soil breaking down due to the extra water. The steeper the slope, the more likely one of these processes becomes.

Landslides and *slumping* tend to be more severe, and usually involve the movement of rock as well as soil. They are common where a heavy rock lies above a weaker rock (Fig. 3.5(*c*)). Saturation of the rocks by heavy rain makes landslides and slumping more likely, because the water has a lubricating effect.

35 Weathering and slopes

QUESTION 3.4 Explain why soil creep is likely to be greater on a bare or lightly-grassed slope than on a densely-wooded slope.

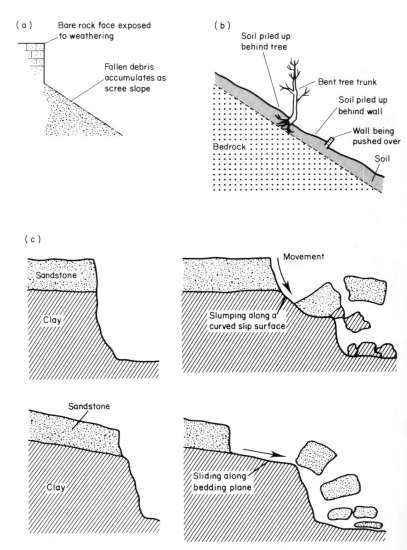

Fig. 3.5 *(a) Scree formation (b) Soil creep (c) Landslides and slumping*

Weathering and slopes 36

QUESTION **3.5** Study Fig. 3.6, which shows a Carboniferous Limestone area in Britain. Identify the features labelled A, B, C, D and E.

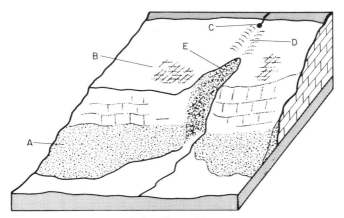

Fig. 3.6 *Carboniferous Limestone scenery*

QUESTION **3.6** Study Fig. 3.7, which shows the position of a settlement.
(a) Describe the position of the settlement.
(b) Give *two* reasons why there is a risk of a landslide or slumping.
(c) Draw a labelled diagram to show what would happen should a landslide or slumping occur.

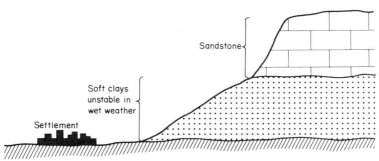

Fig. 3.7

QUESTION **3.7**
(a) Which of the following words describes accumulated rock material at the foot of a bare rock face: (i) Rubble, (ii) Scree, (iii) Creep, (iv) Cement?
(b) Define the term *frost shattering*.

LANDFORMS AND LANDFORM PROCESSES
3.3 Rivers

River systems
The basic features of a river system are shown in Fig. 3.8.

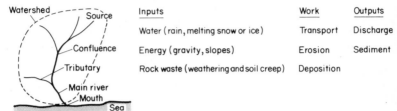

Fig. 3.8 *River system*

A river valley changes along its course. In the *upper part*, the valley is deep and steep-sided (a V shape) as the river is eroding downwards very rapidly – more rapidly than weathering and soil creep can lower the slopes of the valley sides. Downward erosion of the river bed takes place by:

1 **Corrasion or abrasion**, where the river load (boulders and pebbles) wears it away, particularly by pot-holing in which pebbles are swirled around in cracks, widening and deepening them into deep, circular holes.

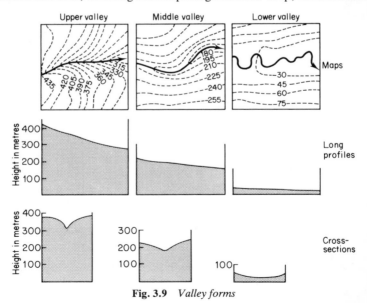

Fig. 3.9 *Valley forms*

Rivers 38

2 **Hydraulic action**, whereby pieces of rock are torn from the river bed if the force of water is great enough.
3 **Solution**, where soluble rocks and minerals are removed.

Most erosion takes place when rivers are high, and the river has the energy to transport its bed load of pebbles and boulders.

Further *downstream*, the valley sides become less steep as the long profile develops a more gentle gradient. Downward erosion here is not as rapid as waste-removal from the slopes. In addition, *lateral* or sideways erosion becomes relatively more important, and the sides of the valley are undercut by the meandering of the river. Very gradually, the valley floor is widened as the meanders cut away at the valley sides. This creates a flood plain (see the Severn flood plain on the Ordnance Survey map extract).

As the river moves, it transports its load, especially in time of flooding. Some material is transported in solution, some is carried in suspension and larger particles are moved by **saltation**, that is, by being rolled along the river bed. As the load is moved, larger pieces are broken into smaller pieces, and rounded and smoothed by contact with each other and the river bed. This process is called **attrition**. Nearer the source there is a greater proportion of larger boulders and pebbles, whereas nearer the mouth there is a greater proportion of fine sands, clays and silts. The flood plain surface consists of transported material in the form of sands, gravels, clay and silt, known collectively as *alluvium*. Rivers wind across the flood plain and occasionally the meanders are cut off, leaving *ox-bow lakes*.

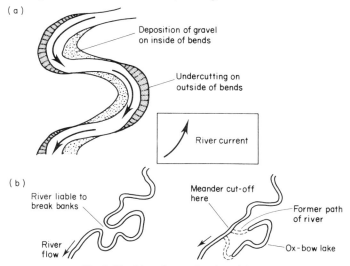

Fig. 3.10 *Meanders and ox-bow lakes*

39 Rivers

Rivers become larger downstream as they are joined by tributaries bringing more water. The flood plain becomes correspondingly wider and the meanders too become wider.

Where a river enters the sea there may be a *delta*, where large volumes of sediment are deposited, building out new land into the sea. In other cases the river enters the sea in a tidal *estuary*, where the daily rise and fall of the tide helps to scour away some of the sediment.

The profile of a river is not smooth, especially in the upper reaches. Hard bands of rock which resist down cutting cause waterfalls and rapids. Waterfalls also occur if there is a fall in sea level and the river starts to cut down to a new *base level*. At the point where a 'new' valley cuts into the old one, there will be a waterfall, or at least a sharp increase in gradient. This down cutting or rejuvenation also results in a new flood plain being cut into the old one. Remnants of the old flood plain are left as *terraces*.

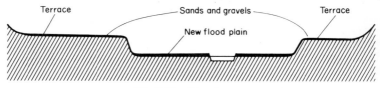

Fig. 3.11   *River terraces*

QUESTION 3.8   Study Fig. 3.12.
(a) Name the features indicated (A–E).
(b) Explain why there is a waterfall at F.
(c) Explain why the settlement at G is prone to flooding.
(d) Describe the formation of feature D.

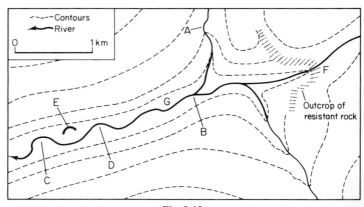

Fig. 3.12

Rivers 40

Larger-scale patterns

On the larger scale, river systems vary considerably, according to:

1 Density of streams and rivers – that is, how much river channel there is in a given area.
2 Pattern of drainage. This may be related to:

(*a*) The structure of the rocks in the area, especially where resistant and less resistant bands of rock are crossed. (Here a rectangular or *trellis* drainage pattern develops, as in southern England).

(*b*) The general form of the area. Where rivers developed on a dome of rocks, they flowed outwards in a *radial* pattern, as in the Lake District and to some extent in Wales.

(*c*) The recent geological history, which has caused rivers to overflow and divert their courses, as with the River Derwent in the Vale of Pickering and most of the rivers along the Welsh/English border. Where there are no special controlling features, *dendritic* drainage patterns are used.

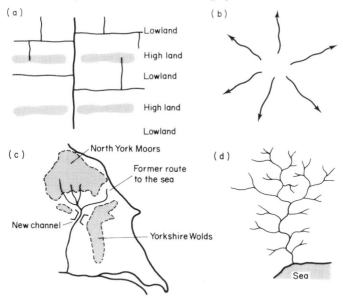

Fig. 3.13 *Drainage patterns: (a) Trellis (b) Radial (c) Drainage diversion in the Vale of Pickering (d) Dendritic*

The type of question on p. 41 is common in examinations. You have to be able to recognise features, explain what processes form them and understand the effects on human geography. You may be tested in short questions, in map-reading questions or in more detailed questions.

41 Rivers

Other features of rivers concerned with **hydrology** are covered in Sections 4.7 and 6.1.

QUESTION 3.9 Study the following showing cross-sections through two waterfalls which have different origins.
(a) Write labels for points A, B and C.
(b) Explain the origin and formation of each waterfall.

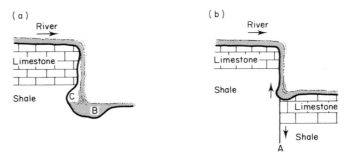

Fig. 3.14 *Waterfalls*

QUESTION 3.10 Study Fig. 3.11, which shows a river flood plain and terraces.
(a) Explain how sands and gravels are deposited across flood plains.
(b) Explain how the terraces were formed.

QUESTION 3.11 Study Fig. 3.15, showing a river passing through a small town, and the various civil engineering works which have been constructed.
(a) What work has been carried out to cut river-bank erosion and flood danger in the town?
(b) What are the possible undesirable results of the engineering work downstream of the town at points A, B and C? Give reasons for your answers.

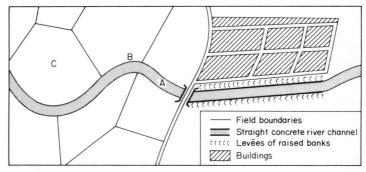

Fig. 3.15

3 LANDFORMS AND LANDFORMS PROCESSES
3.4 Ice

Ice action can be observed in high mountain ranges like the Alps, Himalayas and Rockies as well as the areas covered by huge ice sheets in Greenland and Antarctica.

You may be asked questions about:

1. Glaciers, the features of glaciers and the work done by glaciers.
2. Features of glacial erosion, especially in mountain areas.
3. Features of glacial deposition in upland and especially lowland areas.

Glaciers and their work

Ice caps cover enormous areas and only the highest mountains protrude. At their edges, ice floes *calve* into the sea.

Ice sheets occur on the highest mountain ranges, but are not extensive. *Valley glaciers* often start in those ice sheets. Under the effect of gravity, ice moves downhill very slowly into valleys. Valley glaciers can go well below the normal snowline before melting.

Corrie glaciers form in hollows on the sides of mountains, particularly on the side facing away from the sun.

Ice itself is formed where snowfall in winter is greater than snow melt in summer. Snow accumulates and is compacted by its own weight to form ice.

The features of glaciers largely result from their movement. A valley glacier shows *transverse crevasses*, which occur where the glacier goes over an obstacle, and *longitudinal crevasses*. The glacier contains rock material from three sources:

1. Weathered material that has fallen onto the ice.
2. Material which has been eroded by the ice. This may be loose rock and soil scraped away by the ice, or it may be bedrock removed by the process called *ice-plucking*, as the ice freezes around blocks of rock and tears them away.
3. Material which has been removed by abrasion. With this process, rocks embedded in the ice scrape the bedrock and gradually wear it down, sometimes smoothing the rock and sometimes leaving scratches or *striations*.

Moraines are visible as surface features of glaciers. They are composed of the debris referred to above.

QUESTION **3.12** Study the diagram of a glacier on p. 43.
(a) What sort of glacier is shown?
(b) What is meant by 'zone of accumulation'?
(c) What is meant by 'zone of ablation'?
(d) Explain how the glacier removes rock material from the area.

43 Ice

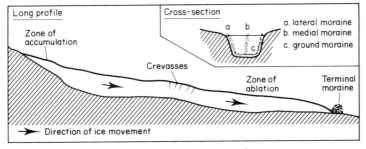

Fig. 3.16 *Features of a valley glacier*

Glacial erosion
Ice sheets generally round and smooth upland areas and in lowland areas scrape the land clear of soil, leaving many hollows in which lakes form when the ice melts.

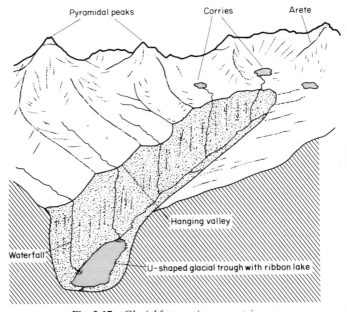

Fig. 3.17 *Glacial features in a mountain area*

Ice 44

A valley glacier has tremendous erosive power due to the enormous volume of ice. The glacier erodes the valley and changes its shape dramatically. It makes the valley *deeper*, *straighter*, *wider* and *steeper-sided*. The V shape of a river valley is changed into the U shape of a glaciated one (a *glacial trough*) as the ice cuts away at the sides and bottom of the valley. The spurs and other obstructions are also cut away forming *truncated spurs*. Side valleys with small glaciers have their lower sections cut away as the main glacier cuts downwards more rapidly. This leaves the side valleys as *hanging valleys*, with a steep fall to the main valley floor, often marked by a waterfall after the ice melts.

A glacial trough may not have a flat bottom, however. In places, the ice may have eroded hollows along the floor of the valley. This could be because the rock is less resistant or because the glacier is able to carry out more erosion, perhaps because a tributary glacier joins the main valley glacier. This would give extra erosive power. These hollows or *rock basins* produce *ribbon lakes* after the ice melts.

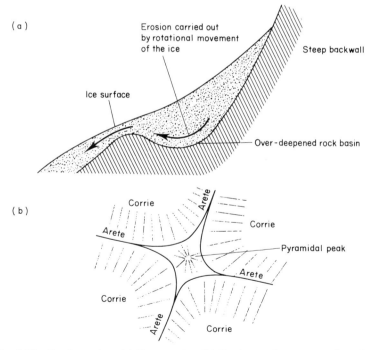

Fig. 3.18 *Corrie erosion: (a) Cross-section (b) Corries eroding a mountain – plan view*

45 Ice

In the time since glaciation ended, the scenery has been modified. Lakes have been partially filled by the debris of streams, building out deltas and alluvial fans. Weathering of the valley sides has produced large screes in many places.

At the heads of valleys and on the sides of mountains above the valleys, corrie glaciers erode bowl-shaped hollows or *corries*. These have a steep back wall and a rock lip often holding back a small lake. Erosion occurs by rotational movement of the ice (Fig. 3.18). Two corries side by side cut back and when their back walls intersect a sharp ridge or *arête* is formed. Where several corries are cutting back into a mountain from all sides, they eventually produce a *pyramidal peak*.

QUESTION 3.13 Examine the map below and identify the glacial features labelled A to F. Choose one feature and explain how it was formed.

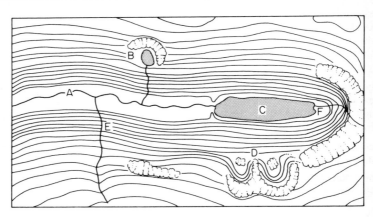

Fig. 3.19

Glacial deposition

When the ice melts, the rock debris within the ice is deposited in a number of ways. Ground moraine is hummocky material which is a mixture of boulder clay (crushed or powdered rock) and larger rocks, and is spread widely. At the snout of the glacier (the furthest point reached by the ice before it melts) is the terminal moraine, a ridge of debris marking the furthest extent of the ice. A lake often forms behind the terminal moraine, but once it overflows it will drain away very rapidly. The lateral moraines at the sides of valley glaciers are left as small ridges of debris.

Erratic boulders are common features in glaciated areas. These are often

Ice 46

found enormous distances from their source, and provide evidence of the direction and distance ice has travelled.

Drumlins are depositional features, where the moraine is arranged in a series of long parallel ridges.

Beyond the furthest extent of the ice, the effect of glaciation is still felt. Meltwaters distribute sands and gravels over a wide area, called the *outwash plain*.

Effects of glaciation

Scenery and tourism The spectacular and dramatic features of glaciation increase the scenic attractions of mountain areas, especially where there are glaciers still present.

Removal of soil In many areas – especially the lowland areas – eroded by ice, very little soil remained, and often the only cultivable areas are the alluvial deposits on the edges of lakes.

Drainage The pattern of drainage in lowland and upland areas is often disrupted. Fig. 3.13(*c*) shows how the River Derwent was diverted from its short, direct route to the sea to a much longer indirect route.

Examination questions may be based on map extracts as well as photographs, sketches or diagrams. You will be expected to identify features and explain how they were formed, as well as considering the effects on human activities. This topic can be covered in all the different types of questions, from multiple-choice to essay questions.

QUESTION 3.14 Draw a diagram to show the hydrological cycle in an area of glaciers.

QUESTION 3.15 The sketch below shows an area before glaciation. What changes will result from glaciation by valley glaciers?

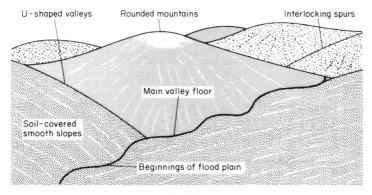

Fig. 3.20

LANDFORMS AND LANDFORM PROCESSES
3.5 Coasts

Coastal landforms can be studied at different *scales*:

1 Small scale, studying individual features along a short stretch of coastline (this is commonly used for examination questions).
2 Medium scale, where features on a regional scale can be observed.
3 Large scale, comparing coastlines on a national or even continental scale.

Processes
The sea erodes, transports and deposits material, forming both erosional and depositional features. In addition, the wind transports sand and deposits sand dunes.

Erosion is a combination of:

Hydraulic action, as the sheer force of the waves beating against rocks in storms forces water into cracks and breaks the rock apart.
Corrasion, as the rock material is thrown at the cliff by the waves and gradually wears away the rock of the cliff.
Attrition, where rock material is broken down further by wave action and rounded and smoothed to form pebbles and sand.

Transport is carried out by wave action moving material back and forth, but movement along the coast is a result of longshore drift.

Deposition is not really a permanent process, because most deposits are in a state of transit.

The sea carries out its actions by wave power. The size and force of waves depends on the *fetch*, that is the distance of open water across which waves pass before hitting land. The greater the fetch, the greater the wave power.

Landforms
Erosion by waves undercuts the base of a cliff, which eventually collapses (see Fig. 3.21). The process continues and results in a wave-cut platform at the base of the cliff. The eroded material may be broken up and deposited as pebbles and sand to cover the wave-cut platform. More resistant rocks often form headlands, and the erosion of the headland produces a sequence of features: weaknesses in the rock are exploited by the erosive power of the waves, and caves develop; eventually, these cut right through the headland to form an *arch*; further erosion leads to the collapse of the arch leaving a *stack*; this is worn down to leave a *stump* or *residual stack* (see Fig. 3.22).

Coasts 48

Differential erosion is an important idea in explaining the development of these features. Weaknesses in the rock, like joint planes or faults, are eroded quickly. On a larger scale, bays and headlands along a coastline may reflect alternating harder and softer rocks.

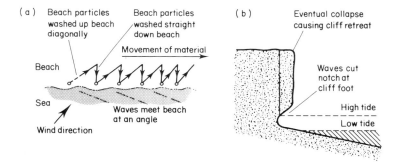

Fig. 3.21 *(a) Longshore drift (b) Cliff erosion*

QUESTION 3.16 Study Fig. 3.22(*b*).
(*a*) Explain the development of the bays and headlands shown on the map.
(*b*) Describe the likely sequence of features that will be formed as the headlands themselves are eroded.

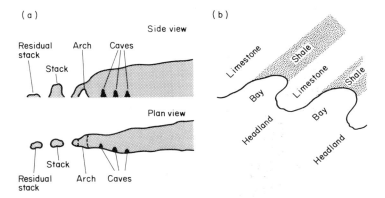

Fig. 3.22 *(a) Erosional features on the headlands (b) Bays and headlands*

Coasts

Depositional features are produced by wave action, too. Beach deposits at the heads of bays are thrown up by *constructive waves* (as opposed to *destructive waves*, which sweep sand away). Where waves strike the beach obliquely, *longshore drift* moves material along the coast and where the mouth of a river occurs a *spit* may be formed, sometimes blocking the exit of the river for many kilometres. Occasionally, a spit will join an island to the mainland, in which case it is called a *tombolo*. Spits also develop where the coast turns away, and then the spit grows in a curving fashion out into the sea.

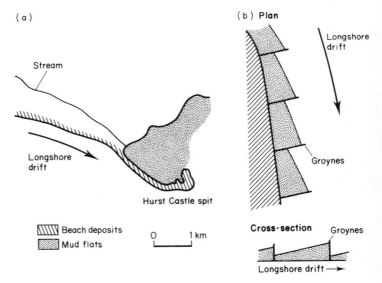

Fig. 3.23 *(a) Spit (b) Beach groynes*

Human interference with coastal processes and landforms has been concerned with:

1 Reducing longshore drift by building *groynes*. Unfortunately, although this keeps sand and shingle on the beaches, groynes deprive places further along the coast of sand. The result is greater erosion of the coast in these places, because the base of the cliff is no longer protected by sand. In several places the result has been much faster cliff erosion, causing houses to fall into the sea and putting whole settlements at risk.

2 Building of sea walls and reinforced barriers to try and reduce or prevent erosion. This is not always successful.

QUESTION 3.17 Which of the following is a result of marine erosion: (a) spit, (b) beach deposits, (c) terraces, (d) arch, (e) moraine?

Larger-scale studies

Particular coastlines have their own characteristics. Areas with generally resistant rocks and high land have cliff coastlines with a great variety of erosional features. Areas with less resistant rocks and lower land tend to have a predominance of depositional features, partly because the more rapid erosion of softer rocks provides more beach material for transport.

In many areas, the coastline has been submerged by a rise in sea level. This is usually the result of the general rise in sea level following the melting of the ice sheets at the end of the Ice Age. Former river valleys have had their lower reaches submerged and made tidal. Such features are called *rias*. Where a glaciated valley has been submerged, the feature is called a *fjord*. In Britain, south-west England has rias and north-west Scotland has fjords. The differences are those between glaciated and unglaciated valleys.

On the very large scale, coastlines can be distinguished by their trend. If the mountains and general relief run parallel to the sea there is a smoother coastline, although numerous islands offshore indicate a submerged mountain range. Such a coastline is called a *Pacific coast*. By contrast, an *Atlantic coast* is one where the trend of relief is at right-angles to the coast, making for a much more rugged coastline.

QUESTION 3.18 Explain why ria and fjord coastlines have an abundance of excellent natural harbours.

WEATHER AND CLIMATE
4.1 Elements of weather

The atmosphere is the layer of gases surrounding the earth. Its lower part is called the **troposphere**, and it is in this bottom layer that everything we call weather occurs. Weather is the state of the atmosphere (or the conditions prevailing) at any one time.

Climate is the general set of conditions to be expected, based on averaging recordings made over many years (generally at least 30 years).

Elements and their measurement
The elements which are recorded and measured are:

Element	*Method/Equipment*
precipitation (rain, snow, hail, etc.)	rain gauge, depth of snow
temperature	thermometers
air pressure	barometer/barograph
wind	anemometer/Beaufort scale wind vane
cloud cover and type	by eye
visibility	by eye
humidity	hygrometer/wet and dry bulb thermometers
sunshine	sunshine gauge

Almost all syllabuses test weather and climate in some way, but it is quite usual for the topic to be linked with another, like hydrology, farming or even tourism. Prepare yourself for this by studying your syllabus and past or sample examination papers.

Questions on weather and climate alone are fairly predictable:

1 You may be given a weather map and asked to describe the weather at particular places, and to describe and explain how it might change in the next few hours. This could be a full question or it could be split into a number of short ones.

2 You may be asked to describe and explain the conditions under which certain types of weather occur, such as frosts, fogs, rainfall over mountains or gale-force winds. Three or four of these would make a full question.

3 In climate questions, you are frequently given climate maps, graphs or statistics for various places. You are asked to describe the climates and to explain the differences between two or more of them.

4 WEATHER AND CLIMATE
4.2 Temperature

The source of heat is the sun, but the actual temperature on the earth's surface is affected by the following factors:

1 Latitude or distance north or south of the equator The sun's heat is more intense in equatorial regions than further away. As a result, equatorial regions receive more heat and have higher temperatures than places in high or middle latitudes.

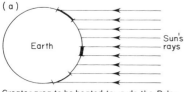

Greater area to be heated towards the Poles than near the Equator, so heating is more intense in lower latitudes

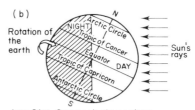

June 21st - Summer in the northern hemisphere

Fig. 4.1 *Variations in intensity of heating (a) Distance from the equator (b) Seasons*

2 Seasons In middle and high latitudes, the length of daylight varies from season to season. Temperatures are therefore higher in summer when there are more hours of daylight and the sun is higher in the sky, giving more intense heating than in winter.

3 Land and sea The sea has a moderating effect on temperature, reducing summer temperatures and raising winter temperatures. As a result, places near the sea have more *moderate temperature ranges* and places distant from the sea have more *extreme temperature ranges*. This effect of increasingly extreme temperature ranges is known as *continentality*. It is particularly marked in continental areas in middle latitudes.

4 Ocean currents These modify temperatures quite considerably. Ocean currents moving towards the poles from tropical regions bring warmer conditions into middle and high latitudes than would otherwise be expected. By contrast, cold currents from polar regions serve to keep temperatures down. The contrasts are marked on opposite sides of each ocean basin.

5 Altitude Temperature decreases by 1°C per 150 metres, so mountain ranges have their own temperature levels, which are quite distinct from the surrounding lowland areas.

6 Cloudiness Areas with more cloud receive less sunshine. However, clouds prevent heat loss by radiation (by reflecting heat back to earth).

QUESTION **4.1** The figures below show the mean January temperatures for four locations in middle latitudes. Which is a coastal location?
a) 8°C, (b) −10°C, (c) 1°C, (d) −3°C.

WEATHER AND CLIMATE
4.3 Rainfall

4

There are three types of rainfall:

1 *Convectional rain* occurs when heated air rises.
2 *Relief rain* occurs when air is forced to rise over high land.
3 *Frontal rain* occurs where warmer air is forced to rise over colder air.

In each case, the air cools due to expansion as it rises. Eventually, it cools to the temperature at which it becomes *saturated*. At this temperature (*dew point*) *condensation* begins to take place. This results in the formation of cloud and, as water droplets are enlarged, *rainfall* follows. The water will, of course, fall as snow or hail under different temperature conditions.

Answers to questions on this topic will need labelled diagrams.

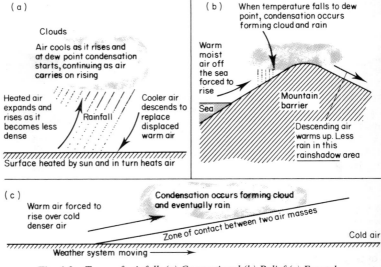

Fig. 4.2 *Types of rainfall: (a) Convectional (b) Relief (c) Frontal*

Most of Britain's rain is frontal, but the amount of rain falling on high land is increased by relief rain, especially in the mountainous west. Convectional rain occurs largely in summer months, especially in the south-east.

Fogs are the result of condensation of water vapour at ground level. *Hill fog* is simply cloud meeting high ground. The two kinds of 'real' fog are:

1 *Radiation fog*, which occurs especially after cold, clear nights when heat is radiated from the earth's surface. The cool land surface then cools the air in contact with it. When its temperature reaches dew point, condensation

Rainfall

occurs and the result is mist or fog. Radiation fogs are particularly common under anticyclonic conditions in autumn and winter.

2 *Advection fog* occurs where warm moist air passes over a cool surface, which may be land or sea. For example, the Grand Banks area of the North Atlantic off Newfoundland has frequent fogs, as the warm moist air from over the Gulf Stream drifts over the cold waters of the Labrador Current.

Considering rainfall as a climatic feature, there are some general statements that should be noted:

1 Rainfall decreases with distance from the sea.
2 Upland areas have heavier rainfall than lowland areas.
3 Inland areas tend to have heavier rain in summer due to convection.

QUESTION 4.2 Study the maps below, showing the distribution of rainfall and high land in the British Isles (Fig. 4.3). Describe and explain the relationship between rainfall and relief.

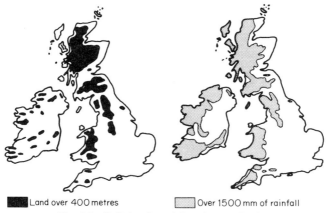

Fig. 4.3 *Relief and rainfall in the British Isles*

QUESTION 4.3 Compare and contrast the rainfall at Bergen and Algiers (Fig. 4.4).

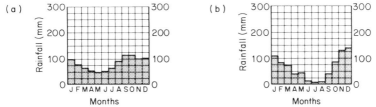

Fig. 4.4 *Rainfall (a) Bergen, Norway (b) Algiers, Algeria*

WEATHER AND CLIMATE
4.4 Climate

4

Climate takes account of all the elements listed in Section 4.1, but for most purposes, rainfall and temperature *regimes* (or annual patterns) are most important. The combination of particular rainfall and temperature regimes results in different climates. Fig. 4.5 shows what you should notice about each climate and what to look for when describing or identifying a climate.

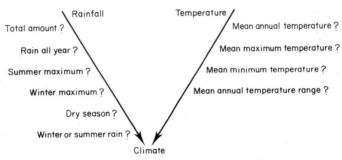

Fig. 4.5 *Steps in describing and identifying climates*

Study the climate graphs for different regions in an atlas and compare the information you derive from the graphs with that in Fig. 4.7 (p. 56) and the Table on page 57. Make sure that you can explain the differences between climates as well as describing them. Note that the names used for climatic types vary according to the climatic classification. Most names are self-explanatory, even if they are not the ones you have used.

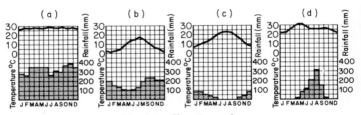

Fig. 4.6 *Climatic graphs*

QUESTION 4.4 Study the set of climate graphs above.
(a) Describe the climate shown by each graph.
(b) Using the table on p. 57 as a source, match the names in this list to the climates you describe: West Coast, Mediterranean, Tropical wet summer, Equatorial. Give reasons for your choices.
(c) Explain why agriculture faces problems in the region shown by graph (d), despite the fact that total rainfall is as high as in lowland Britain.

Climate 56

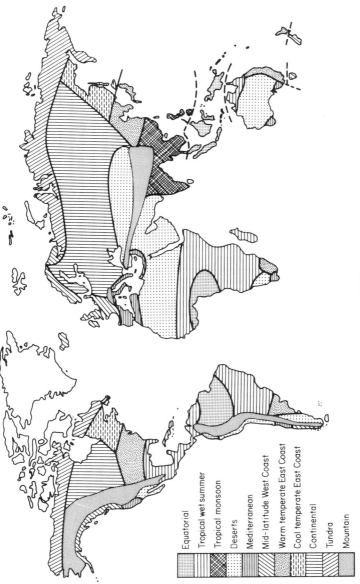

Fig. 4.7 *World climate zones*

Examples of climates

Climatic type	Example	Description
Tropical		
Equatorial	Para, Brasil (1°S)	Rain all year (2437 mm), ranging from maximum of 357 mm in February to minimum of 66 mm in November. Temperatures high, only varying between 25°C and 26·7°C
Tropical wet summers	Kano, Nigeria (12°N)	Total rainfall of 861 mm, rising gradually from 68 mm in May to a maximum of 310 mm in August before decreasing rapidly. Temperatures range from 21·7°C in January to 30·6°C in May
Tropical monsoon	Bombay, India (19°N)	Total rainfall of 1800 mm, almost all during June to September. Temperatures are high, ranging from 23·9°C in January to 29·4°C in May. Rainy season starts suddenly
Hot desert	Aswan, Egypt (24°N)	No rainfall (unusual as most desert places have some). Temperatures vary from 15·6°C in January to 33·3°C in July
Temperate		
Mediterranean	Algiers, Algeria (37°N)	Total rainfall of 762 mm, most falling November to January. Summers dry and hot (25·6°C) and winter warm (12·2°C)
Mid-latitude West coast	Brest, France (48°N)	Rain all year (861 mm), heaviest in winter. Maximum temperature 17·8°C, minimum 7·2°C, small range of 10·6°C
Warm temperate East coast	Raleigh, N. Carolina, USA (36°N)	Rain all year but summer maximum. Heavy rainfall of 1142 mm. Mild winters (5°C in January) and hot, humid summers (26·1°C)
Cool temperate East coast	Vladivostock, USSR (43°N)	Rains falls mainly May to October, totalling 606 mm. Extreme temperature range (−13·9°C in January and 20·6°C in July)
Continental	Prince Albert, Saskatchewan, Canada (53°N)	Rain all year (snow in winter) with summer maximum. Total 404 mm. Extreme temperature range from −20°C in January to 17·2°C in July
Cold		
Tundra	Thule, Greenland (76°N)	Negligible rainfall (88 mm). Minimum temperature −29·4°C, maximum temperature 4·4°C. Only three months above 0°C

4 WEATHER AND CLIMATE
4.5 Special climates

Mountains These have their own climates, which are cooler because of the altitude. Temperatures decrease with altitude at a rate of 1°C per 150 metres. This is known as the lapse rate. Note, however, that the temperature pattern throughout the year is like that of nearby lowlands but lower. The actual mean monthly temperatures will depend on altitude. In the highest mountain ranges, it is possible to go through a whole series of climatic zones from Equatorial to Arctic.

More locally, there are important features which affect human activities. Where valleys in mid-latitudes run from east to west, the south-facing side receives more sunshine than the other side, especially in winter, and is therefore a more favourable location for settlement. This is very noticeable in parts of the Alps.

Frosts These are most common in hollows or at the foot of a slope. On clear winter nights, radiation of heat reduces air temperatures near the ground and cold air flows down slopes to collect in hollows or on lower ground. These places have more frequent and heavier frosts than places on the slopes above. This condition, in which ground temperatures are lower than the air above, is called a *temperature inversion*.

Urban areas These have their own special climates, too. The most important feature is the *urban heat island*. Buildings and vehicles, for example, give off heat and the temperature of urban areas can be several degrees higher than that of the surrounding countryside. However, increased atmospheric pollution can lead to a great number of fogs.

Examination questions either ask for an explanation of these features as one of three or four smaller sections, or give you information on a particular place and ask you to describe and explain the conditions.

QUESTION 4.5 Study the diagram below, showing a cross-section of a valley in Britain in mid-morning in January.
(a) Compare and contrast the weather conditions at A and B.
(b) Explain why there is fog lying in the valley bottom.

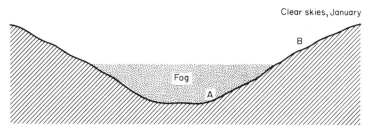

Fig. 4.8

WEATHER AND CLIMATE
4.6 Weather systems

Most of this section concerns the weather of the British Isles, but the principles can be applied to almost any mid-latitude area.

One reason why the weather of the British Isles is so varied is that it is affected by *air masses* from different sources (as shown very simply in Fig. 4.9). Clearly, the *temperature* and *humidity* of the air is very important in affecting weather.

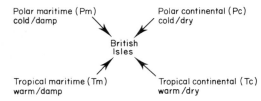

Fig. 4.9 *Air masses affecting the weather of the British Isles*

The second reason why our weather is so varied is that the country is affected by different pressure systems, often in rapid succession. There are two basic pressure systems: high and low. These are shown on a map by **isobars** (lines joining places of equal pressure). The essential difference is that areas of high pressure or **anticyclones** bring stable, generally dry weather conditions. Low-pressure systems or **depressions**, on the other hand, bring unstable and often wet and windy conditions. This is because within an anticyclone air is descending, whereas in a depression air is rising. Winds blowing towards or away from the centres of these systems are deflected to the right due to the earth's rotation. The result is that in an anticyclone winds blow clockwise, parallel to the isobars, and in a depression they blow anti-clockwise. (Note that this is reversed in the southern hemisphere.) In depressions, isobars are generally closer together than in anticyclones. This means that there is a steeper *pressure gradient* and is the reason for the stronger winds in a depression. Isobars close together mean strong winds and isobars far apart mean light winds.

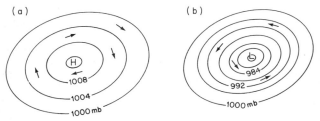

Fig. 4.10 *(a) High pressure – anticyclone (b) Low pressure – depression*

Weather systems 60

Anticyclones over Britain are generally associated with Pc air masses in winter and Tc air masses in summer (see Fig. 4.9), and the temperature conditions obviously reflect their sources. Depressions generally develop along the boundary or *front* between Pm and Tm air masses. In between depressions are *ridges* of high pressure, which bring shorter periods of stable weather.

Frontal depressions pass over the country frequently. They usually track further south in the autumn and winter bringing more changeable and wetter weather, but in any particular year this general statement may not apply.

A frontal depression brings with it a particular sequence of weather. The changes are shown on the cross-section and map (Fig. 4.11). Note the differences in temperature between the warm and cold sectors of the depression, the changes in wind direction, the variations in cloud cover and the location of rain belts.

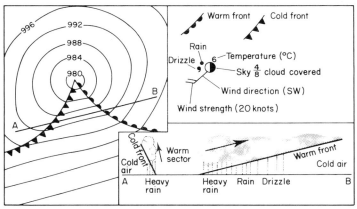

Fig. 4.11 *Frontal depression*

As a frontal depression moves across the country from west to east, so the weather changes. Imagine the cross-section standing upright and being moved across a map of the country. For your own area, work out how the weather would change, first as the warm front approached and passed over and then later as the cold front passed. You should note (*a*) what the weather is like in the cold sector in front of the warm front, (*b*) what it is like in the warm sector, and (*c*) what it is like in the cold sector behind the cold front.

Anticyclones generally bring clear skies. In summer this produces a spell of hot, dry weather, but in winter the clear skies increase heat loss by radiation. It is common for very low surface temperatures to produce

61 Weather systems

temperature inversions, giving frosts and sometimes persistent fogs at the surface. Anticyclones tend to persist for days, weeks or even months in exceptional circumstances. The ridges of high pressure referred to earlier bring only short periods of stable fine weather between the changeable conditions of depressions.

QUESTION 4.6 Study the weather map of Western Europe below
(a) (i) Describe the weather conditions at stations K and M.
(ii) Label a depression and an anticyclone.
(iii) Explain why it is colder at station L then at station N.
(b) Forecast the weather changes in northern Scotland over the next twelve hours.

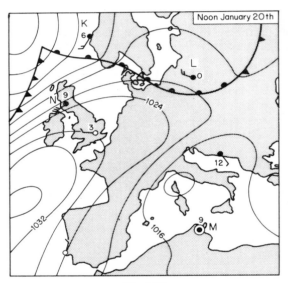

Fig. 4.12

Make a point of watching the BBC television weather forecasts regularly, as this will help you to understand the system.

4 WEATHER AND CLIMATE
4.7 Hydrological cycle and hydrology

The **hydrological cycle** summarises the way that water moves through the environment. All the key words are shown in Fig. 4.13. You must be familiar with them and be able to define them.

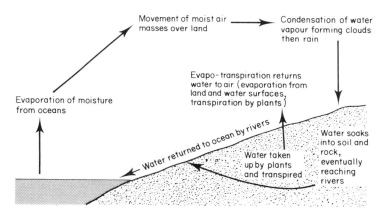

Fig. 4.13 *The hydrological cycle*

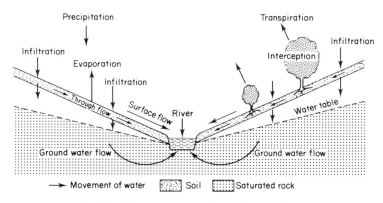

Fig. 4.14 *Water movement within a drainage basin*

QUESTION 4.7 Describe, in words, the movement of water in the biosphere. (This question incorporates a test of your knowledge of *terms* as well as your understanding of the whole *system*.)

63 Hydrological cycle and hydrology

The more local effects of rainfall and water movements are usually studied in the context of individual *river basins*. This local level of study is important as it links up with studies of flooding and soil erosion, which are both affected by the characteristics of the river basin. Within a river basin, water that falls as rain can follow several different routes through the system before leaving the basin via the river. Note the terms used in Fig. 4.14 (p. 62).

The rate of movement of the water affects how quickly or how slowly the river will rise, and later fall, after it starts to rain. The two graphs in Fig. 4.15 are examples of *storm hydrographs*.

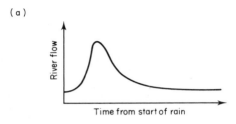

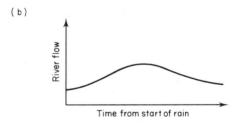

Fig. 4.15 *Storm hydrographs*

The table on page 64 summarises how local drainage basin features affect water movements.

Hydrological cycle and hydrology

Local features and water movement

Local features	Hydrological effect
Drainage density	The more streams, the shorter the distance water has to travel to reach a stream
Slopes	Water will run down steeper slopes and is less likely to infiltrate into the soil
Rocks	Permeable rocks will absorb water, whereas impermeable rocks ensure that water stays on the surface
Soils	Deep porous soils will absorb more water than shallow heavy soils
Vegetation	Vegetation intercepts rainfall and allows water to reach the soil more slowly, increasing the chance of it soaking in instead of running off
Temperature	High temperatures will increase water loss by evaporation and transpiration
Previous weather	If previous conditions have been very wet the soil could be saturated and unable to absorb water. The opposite would be true after a dry spell

The following question is a fairly typical one, but the information may be presented as sets of maps or sketches instead of descriptions, and an Ordnance Survey map may be used as the source of some information.

QUESTION 4.8
(a) Match the two descriptions to the two storm hydrographs in Fig. 4.15:
Description 1 The drainage basin is in a mountainous area of hard, impervious rocks with thin soils and a considerable amount of bare rock. Rough grass is the only vegetation.
Description 2 The drainage basin is in an undulating area of relatively soft, permeable rocks with deep soils. Much of the area is covered by woodland.
(b) What would be the hydrological effects if the area in description 2 became the site of a new town?

On an annual basis, there is a close relationship between river flow and climate – especially the annual distribution of rainfall – and in some areas, the time of snow melt. Note that a river can rise in a region with heavy rains in one season and flow through a region with a different rainfall regime or no rain at all. The River Nile is an example of this, with seasonal rises and falls linked to its source regions.

NATURAL ENVIRONMENTS
5.1 Natural regions

The importance of natural environments varies from one syllabus to another. However, you ought to know in general terms where different kinds of natural environment are found and what their characteristics are. This is relevant to many other topics, such as population distribution, agriculture and the exploitation of resources.

Natural regions are based on the natural vegetation, even though in many parts of the world there is no longer any truly 'natural' vegetation. The predominant type of vegetation is one level of **ecosystem** (Fig. 5.1). Ecosystems involve the complete set of relationships between living things and the non-living environment in an area. At the scale of natural regions the most important single factor is climate, but Fig. 5.2 gives a more complete picture of the factors involved in an ecosystem.

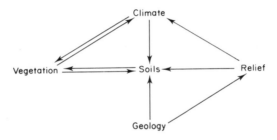

Fig. 5.1 *Ecosystems – linkages in natural environments*

You can see from this map that the relationships involved can be quite complex. Here are two examples:

1 Tropical rain forests High temperatures all year round ensure that plants can grow throughout the year, and rainfall throughout the year assures adequate water supply. In turn, plants can grow with no seasonal effects. Dead plants decompose rapidly in hot moist conditions and their nutrients are taken up very quickly by the roots of living plants. The soils themselves contain little nutrient, since heavy rains wash it out. Provided there is no interference, this ecosystem will persist.

Natural regions 66

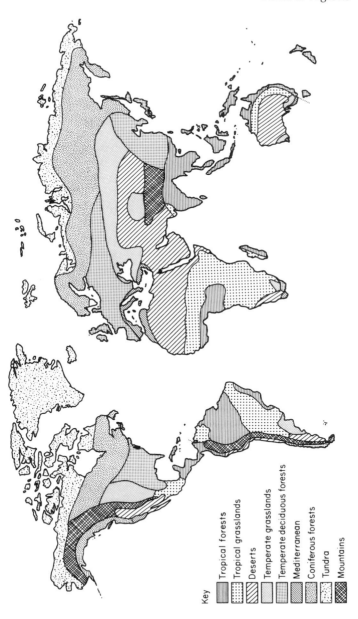

Fig. 5.2 *Natural vegetation zones*

67 Natural regions

2 Northern coniferous forests In northern latitudes, winters are long and very cold. Coniferous trees have developed so that they can conserve food in their root systems and store water by means of a thick cuticle on their leaves or needles. They are adapted to living on poor soils which are heavily leached by the water released in the spring thaw. The relatively short summer means that growth is slow.

Typical questions might ask you to explain why there are seasonal variations within one region but not within another. You might be asked why the natural vegetation changes along the line of a transect, perhaps from the equator to the Tropic of Cancer in West Africa or from the Arctic Circle to 40°N in the USSR.

Since detailed knowledge of natural regions is rarely required, you should refer to other books if your syllabus demands detailed study. Take care, in that case, to find out which regions you have to study and to make notes on at least one regional example of each.

Types of natural vegetation

Tropical forests The tropical rain forests of the Amazon, Zaire Basin and West Africa and South-East Asia, together with the monsoon forests, are notable for the enormous variety of species. The trees form layers at different heights and shade out most light from the ground.

Tropical grasslands These are located between the tropical forests and the deserts and are also called *savanna*. The type of grass varies according to rainfall as does the number of trees. Towards the boundary with the tropical forest are more trees, whereas towards the desert even grass becomes sparse. Trees are adapted to withstand drought.

Deserts On the fringes of the deserts are highly specialised plants which can withstand drought. Within deserts, the only significant vegetation is at oases and along rivers.

Temperate grasslands Called *steppe*, *prairie* and *pampas*, these are now the world's major wheat-producing areas, as well as being used for grazing. The low rainfall, high summer evaporation rates and cold winters restrict tree growth.

Temperate deciduous forest Mostly cleared for cultivation, especially in Europe. They occur where there is more rainfall and less extreme temperature conditions. There is a much more restricted range of species than in tropical forests.

Mediterranean forest Very little real forest remains, having been cleared by burning and grazing. Trees are adapted to overcome the summer drought as is the scrub vegetation which now covers much of the land not used for cultivation.

Coniferous forest The Taiga or northern coniferous forest has conifers adapted to withstand the cold winters. There is a very limited range of species; often there are large stands of one species.

Tundra Extreme cold and short growing season eliminate all but the hardiest plants. The landscape is treeless with a very limited range of common plants. Alpine regions have similar but more varied plants.

5

NATURAL ENVIRONMENTS
5.2 Hostile environments

Environments can be regarded as hostile for a variety of reasons, but on the world scale it is generally climatic factors that are most important (although with mountains, the relief element is important in itself, apart from its effect on climate).

Cold regions

These limit human activity by their short growing season, which affects farming, and by the need to import food, housing and heating fuel for any population except those few living by traditional activities, such as hunting and fishing. These regions can support only a very small population living by these means. Where there are greater numbers of people, it is usually due to the presence of mineral or fuel resources, whose high value makes the extra costs of developing the resource economically worthwhile. Even so, there are other problems. *Permafrost* is the term used in tundra regions to describe the way the soil and often bedrock is permanently frozen, except for the thawing of a thin surface layer in the short summer. This results in waterlogging of the soils and the creation of large areas of marsh, creating movement problems. In winter, extreme cold combined with frequent blizzards is the obvious hazard. Buildings, roads, railways and pipelines all face difficulties because of the thawing and freezing of the surface. Buildings have subsidence problems. Techniques have been developed to deal with the problems, but they are expensive.

Deserts

These are hostile due to their lack of water rather than their high temperatures, although the latter does increase the rate of evaporation of any water that there is. Habitation is only found where there is water, either present naturally or brought in by tanker or pipeline. Water occurs naturally in springs or where bore holes tap underground water. Here oases develop. In traditional oases, the system of cultivation is adapted to take account of problems of evaporation and the limited supply of water. More modern approaches often result in problems of salinisation, as shown in Section 6.2.

It does rain in deserts and this is the source, strangely, of another hazard. With no vegetation, the rain runs off along dry valleys or *wadis* and causes an enormous amount of erosion. Roads crossing wadis are washed away, settlements in their path suffer damage and people can be drowned.

Mountains

These make a third hostile environment. They are barriers to movement and even the most modern roads or railways are susceptible to avalanche and landslide damage. In the steeper, more unstable mountain areas, landslides are an ever-present hazard. Soils and climate limit agricultural development. Where other factors are favourable, as in parts of Indonesia and the Philippines, terracing can take agriculture up to high altitudes. Throughout most of the world, however, mountains present limitations and ultimately, altitude and lack of oxygen prohibit human occupation.

Other aspects of hostile environments are considered in Sections 6.1 and 6.2. Note that locally, the processes of erosion can cause dangers; earthquakes and volcanoes, too, should be considered under this heading.

Questions on this topic often make up part of another question. Thus, a question about semi-arid areas might ask about the difficulties presented to farmers by the environment.

QUESTION 5.1 What are the problems of farming in areas of steep mountains in (a) a tropical area with heavy rainfall, and (b) an area in middle latitudes.

QUESTION 5.2 Fig. 5.3 shows a dam and lake in a mountainous area subject to earth tremors. What are the possible dangers?

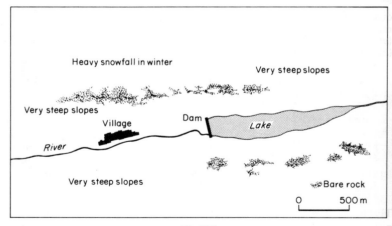

Fig. 5.3

6

PEOPLE AND NATURE
6.1 Hydrology, flooding and erosion

Most human activities interfere with the hydrological cycle in some way. The different circumstances affecting local water movements listed in Section 4.7 can be extended to include changes brought about by people. Some of the changes possible are:

1 Clearing woodland for agriculture.
2 Clearing hedgerows and leaving huge fields bare for much of the year.
3 Putting drainage channels into fields.
4 Building on the land: houses, factories, roads, drains, sewers, streams in culverts.

In all cases, the effect is to speed up the movement of water through the local system, so that water falling as rain reaches the river more quickly. This is done by: reducing interception, so allowing water to reach the surface more quickly; reducing soil infiltration capacity, partly by rain impact on bare soil and partly by machinery compacting the soil; increasing surface run-off by reducing infiltration; speeding up the movement of soil water; draining water from the surface to artificial channels.

In questions on this topic, *careful observation* of the information you are given is vital if you are to make full use of it.

QUESTION **6.1** Study the two sketches below, showing a drainage basin with a reservoir in a mountain area which receives heavy rainfall.
(a) What will be the effect on water movement of clearing the forest?
(b) What will be the effect on the reservoir itself?

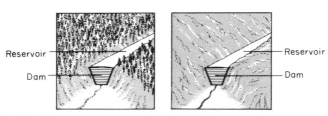

Fig. 6.1 *Mountain reservoir and catchment before and after deforestation*

This question does not confine itself to hydrology. It includes landforming processes or erosional processes. This is common. It is also common to use sets of sketches like the ones in Fig. 6.1, which show changed circumstances; your task is to reason out the effects of the changes. The same points would be covered by this next question, based on the Ordnance Survey map extract.

QUESTION **6.2** What would be the hydrological and erosional effects if the woodland was cleared in square 2506 and 2606?

PEOPLE AND NATURE
6.2 Soil erosion and desertification   6

The removal of soil by erosion is a major problem in many parts of the world. It is particularly severe in areas with light and unreliable rainfall. It is, however, a problem even in humid areas like Britain.

Soil is a vitally important resource. It consists of:

1 Mineral materials in the form of sand, clay and silt in varying proportions.
2 Organic matter or *humus*, formed from the breakdown of plant material.
3 Water.
4 Air.

The type of soil depends on the parent material from which it is derived, the climate and the type of vegetation. Where the land is used for agriculture, the system of farming is important rather than the original vegetation.

A fertile soil will contain plant nutrients like nitrates, phosphates and potassium.

Soil erosion occurs by:

1 Wind blowing away fine particles of soil.
2 Water
   (*a*) Sheet erosion, where water rushes in a sheet across the surface removing loose material.
   (*b*) Gully erosion, where water is concentrated and cuts deep gullies. As well as removing soil, often on a large scale, gullying lowers the water table so that even deep-rooted plants cannot reach the water supply.

The causes of soil erosion are:

1 Overcultivation, especially by growing one crop year after year (**monoculture**). This exhausts the soil and causes it to lose its structure, so that it breaks up into fine particles ready to be blown away.
2 Cultivating steep slopes in areas of heavy rainfall.
3 Removing hedgerows which act as windbreaks.
4 Leaving soil bare for long periods.
5 Overgrazing, causing the roots of grasses to die.
6 Clearing of forest cover.

All of these have effects which make the soil free for wind erosion, or cause hydrological changes which increase surface run-off and in turn erode the soil by sheet or gully erosion.

Soil erosion can be avoided or remedied by:

1 Crop rotation and fertilisers, which return nutrients to the soil and help soil structure.
2 Avoiding ploughing steep slopes, or ploughing parallel to the contours (contours ploughing), or building terraces.
3 Planting hedgerows or rows of tall trees as windbreaks.
4 Planting a ground-cover crop to avoid large areas of bare soil, or strip crop with alternating strips of grass and crop, or inter-crop with alternate rows of crops.
5 Reducing sizes of herds or changing methods of herding cattle.
6 Afforestation.
7 Allowing fallow periods in which soil can recover.

Note that soil is destroyed where eroded material is deposited. It is common for fine alluvial soils in valley bottoms to be covered by coarse deposits eroded from hillsides above.

In arid areas, salinisation is a problem. This is caused when water is drawn up to the surface by high rates of evaporation brought about by high temperatures. As the water evaporates, salts are left behind, and these eventually making the soil useless for cultivation. This can be a severe problem in arid and semi-arid areas, where land is irrigated on a large scale. Expensively reclaimed land is turned into useless salt-encrusted land.

Desertification is a recent term to describe the way some areas are being turned into desert. The areas at greatest risk and suffering the worst damage are on the edges of the world's deserts. These *dry lands* have low and unreliable rainfall, 30 per cent of the world's land surface and a sixth of the world's population (650 million). The causes listed above have a greater effect here because the environment is more fragile and susceptible to damage.

Questions on this topic are of two basic types:

1 Straightforward question asking for the causes of and remedies for soil erosion, in which you usually have to refer to *named examples*. Depending on the wording of the question and your syllabus, you can choose examples from anywhere in the world or from specific areas.
2 An example is presented, and your task is to extract specific facts from the information given, to apply your knowledge in order to explain what is happening and possibly to suggest specific courses of action. This type of question is increasingly common.

QUESTION **6.3** State the causes of soil erosion. Refer to examples you have studied.

QUESTION **6.4** What are the remedies for soil erosion? Refer to examples in North America.

73 Soil erosion and desertification

QUESTION 6.5 The area shown below is at great risk from soil erosion.
(a) Explain why and how the valley sides are likely to suffer from soil erosion.
(b) How is the farmland on the valley floor likely to be damaged as a result of soil erosion on the valley sides?
(c) Suggest the methods that could be used to reduce the risk of soil erosion or to repair the damage of soil erosion.

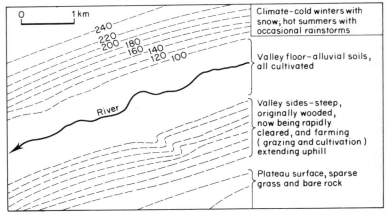

Fig. 6.2

6

PEOPLE AND NATURE
6.3 Pollution

Pollution is a world-wide problem and most syllabuses expect you to have some knowledge of the types and effects. As a topic, it is usually examined as part of a question rather than as a full question.

There are different types of pollution:

Air pollution

This is due to burning of fossil fuels, producing sulphur dioxide, ash and soot; road vehicles and factories producing carbon monoxide, nitrogen oxide, lead and other pollutants.

Locally, these have a greater effect where atmospheric conditions create temperature inversions, trapping pollutants in the bottom layer of the air and creating 'smog' conditions (a combination of fog and smoke). This is less common in Britain now, but in December 1953 a heavy smog in London killed over 4000 people – largely those susceptible to lung complaints.

Water pollution

This is due to:

   effluent from sewage systems
   waste from factories (including chemicals and hot water)
   fertiliser washed from fields

These not only pollute rivers from direct outlets, but can also pollute ground water by percolating downwards and thus they can be found in drinking-water supplies. The effects in rivers vary according to the waste itself, but it is always harmful to the natural environment and often directly harmful to people.

Sea water is polluted in the same way, but with extra hazards caused by oil slicks from oil wells, cleansing of tanks by ships, general leaks as a result of collision and accidents, particularly at ports.

Land pollution

This is due to:

   dereliction
   rubbish tips
   contamination by the dumping of chemicals
   unauthorised dumping of litter

There are innumerable ways by which land is polluted, but all have the effect of making the land useless for any other purpose, in the short term at least.

Many types of pollution affect land, water and air, like radioactive contamination from nuclear explosions or accidents (see Section 13.2).

75　*Pollution*

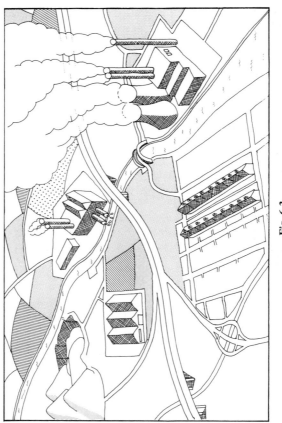

Fig. 6.3

Pollution 76

Noise pollution is also recognised and is found near airports, in the vicinity of busy roads and motorways (which also produce air pollution), and near heavy industry.

Visual pollution is a term used to describe unsightly features.

QUESTION **6.6** Study Fig. 6.3.
(a) What types of pollution are shown? Give one example of each type.
(b) Suggest a possible set of measures to remove, or at least reduce, the effects of the pollution shown.

Remedies for pollution all involve a change of approach. Some of the more dramatic results have come from new laws like the Clean Air Acts of 1956 and 1968 in Britain. Others come from changes in public opinion, sometimes brought about by the publicity generated by particular pressure groups.

One kind of question which is sometimes asked involves role play. You are expected to view the situation through the eyes of particular individuals or organisations, and to give their points of view. Background information is usually given as well as the occupation or position of those concerned.

QUESTION **6.7** The Environmental Services Committee of Washwell District Council has proposed the building of a new sewage works on the edge of the town. Study the map below, showing the town and the location of the proposed sewage works.
(a) Describe the location of the proposed sewage works.
(b) What is likely to be the view of the residents of the area adjoining the works?
(c) What would be the advantages of locating the sewage works at the point marked X?

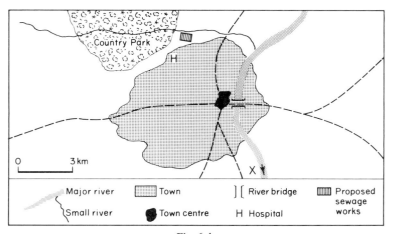

Fig. 6.4

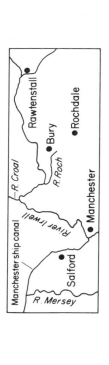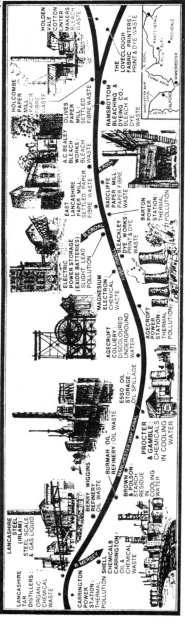

Fig. 6.5

Pollution 78

QUESTION 6.8 Study Fig. 6.5, showing sources of pollution along a stretch of the River Irwell. Improvements in water quality had already been made at the time and there has been much improvement since.
(a) Through which major city does the River Irwell flow?
(b) In what range of hills does the river rise?
(c) What is meant by the term 'water quality'?
(d) Name one source of each of the following types of pollution along the River Irwell: (i) heat, (ii) chemicals, and (iii) solids.
(e) In what ways would the River Irwell have differed from an unpolluted river?
(f) Suggest means by which pollution of the river might have been reduced and ways by which the area could have gained as a result of the cleaning up of the river.

PEOPLE AND NATURE
6.4 Destruction of the natural environment

6

Small-scale environments can be destroyed or damaged by pollution, but on a large scale, a whole variety of processes is involved:

urbanisation, or the spread of cities (see Unit 9)
desertification (see Section 6.2)
pollution of large lakes and inland seas
destruction of tropical rain forests
damage by acid rain
destruction of heathlands and wetlands (usually for use as farmland)

Pollution of large lakes and enclosed seas

Lake Erie in North America was one of the first examples of almost total destruction of a water environment caused by pollution from cities and industries. Two sets of circumstances have saved the lake: (*a*) action to control and limit pollution, (*b*) the decline of heavy industry in the region.

The Mediterranean Sea illustrates an even wider range of factors. Pollution comes from domestic and urban sewage, industry, oil, agricultural fertilisers and pesticides. Some of this pollution comes from well inland and is carried to the sea by large rivers. Action has been taken to start controlling pollution, but it is an urgent matter for a number of reasons: the population of the region is likely to double by the end of the century; more jobs in industry will create more industrial effluent; an increase is forecast on the already vast numbers of tourists visiting the region; the less developed countries surrounding the sea are likely to give priority to jobs and food production rather than pollution control. Failure to control pollution will lead to disease, spread directly or through marine life eaten by people. More importantly, the breakdown of organic pollutants will use up oxygen, the lack of which will kill all marine life.

QUESTION 6.9
(*a*) Explain the link between population increase and increased pollution.
(*b*) Draw a diagram to show how the causes of pollution are linked.

Destruction of tropical rain forests

The tropical rain forests of the world are disappearing at a very rapid rate, something like 20 hectares *per minute*. They are being cleared for three purposes:

1 To provide timber, since the forests are the source of valuable hardwoods.
2 Grazing land for beef cattle is created on land burned and cleared.
3 Shifting cultivation is removing more and more forest because of population pressures.

Destruction of the natural environment

There is no real agreement on which of these is most important, and it does seem to vary from one region to another.

Importance of the rain forests
1 They provide many species of plant now widely cultivated (e.g. rubber and coffee).
2 Unknown species (possibly thousands) may provide medicines, food and genetic diversity to fight disease in cultivated crops.
3 They provide fuel and plantation crops.
4 They protect the land against flooding and silting.
5 Game reserves provide the basis for tourism.

Results of deforestation
1 Soil erosion.
2 Flooding downstream of cleared area.
3 Silting downstream, clogging reservoirs, irrigation systems and covering farmland.
4 Loss of animal species.
5 Loss of plant species.
6 Loss of human habitat.
7 Climatic changes, due to changes in the local and regional hydrological cycle.
8 Global climatic changes, as loss of forest may mean an increase in carbon dioxide in the atmosphere and a rise in temperatures.

Questions on this topic could be short and straightforward, asking for specific statements, or they could be short questions on some information given to you. Another type of question could present statistics of the changes in some countries, ask you to describe these changes and then go on to suggest the possible effects. Yet again, there is the possibility of a straightforward short essay, as part of a full question.

PEOPLE AND NATURE
6.5 Conservation and conflict

6

Many questions on the theme of 'people and the environment' will focus on the conflicts between different uses of the land and conservation. Conservation is to do with the sensible use of resources. If damage is avoided in the way the resources of the environment are used, then they may continue to be used in the future. Conservation does *not* mean that places are preserved unchanged and unused.

There are many different examples of this sort of issue given throughout this book. Here is one example for you to think about:

QUESTION 6.10
Conservation and conflicts in a mountain area
Study the map (opposite) of a mountainous area in Western Europe.
- (a) List all the different ways in which the land is used. (Note that this means the different purposes for which the land is used.)
- (b) Which uses are likely to cause permanent damage to the environment?
- (c) Which uses might enhance the total environment by creating new local environments?
- (d) Which uses conflict with each other? Explain your answer.
- (e) Which uses depend on the area being conserved? Explain your answer.
- (f) There is a proposal to build cable-cars, chair-lifts and drag-lifts for skiing in area A. Suggest reasons for and against this proposal.

These questions add up to much more than a full examination question, but they cover the range of questions you might be asked, whether about a mountain environment or any other kind of area. Remember that the same questions would apply at a very much smaller scale, like the effect of a new colliery on the local landscape. Remember, too, that you should consider all the effects, not just the obvious ones. In question (*f*), for instance, new approach roads and extra accommodation would be needed.

Conservation and conflict 82

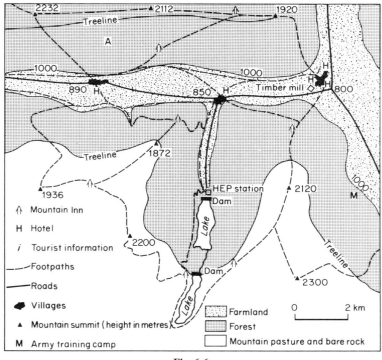

Fig. 6.6

POPULATION
7.1 Population distribution

At all scales, population is distributed unevenly. Population distribution is usually shown on maps in one of two ways:

1 Dot maps, where one dot represents a certain number of people. The pattern made by the dots should give a clear indication of the more crowded and less crowded areas.
2 Density shading maps, where areas are shaded according to the number of people per square kilometre.

It is usual to talk about *densely-populated*, *moderately-populated* and *sparsely-populated* areas, although the middle group tends to include an enormous range of circumstances. Dense populations would have over 100 per km^2, while sparse populations would have less than 1 per km^2 (or less than 5 per km^2). In between would be the wide range of moderate population. Other terms would be *high*, *medium* and *low* population density.

You can use the same terms when talking about world, regional or local population patterns. In the second and third, it would be better to say *relatively* high or low, to emphasise the comparison rather than the actual numbers.

World population distribution

Some areas of the world are virtually uninhabited, and others extremely crowded with 70 per cent of the world's population living on less than 10 per cent of the land surface. The concentrations of population are:

1 South and East Asia.
2 Europe.
3 North-east USA.
4 Lower Nile Valley in Egypt.
5 West Africa.
6 Coastal regions of South America.

These areas have high densities for a variety of different reasons – some physical, some human and economic (including historical reasons). They are able to support large numbers of people, owing to a combination of circumstances: they had the resources from which manufacturing industry could develop; they developed trade; favourable climate and soils supported agriculture; highly productive farming systems were developed; historical reasons saw the settling of large numbers of migrants.

By contrast, the sparsely-populated areas are hostile environments in some way. They are the cold areas of high latitudes and high altitudes, the hot deserts and cold or temperate deserts and the denser and more isolated tropical rain forests. The *physical* factors which explain their low density

Population distribution 84

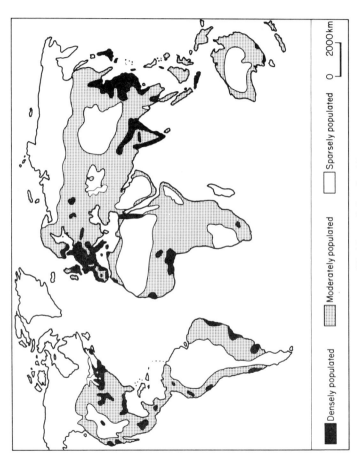

Fig. 7.1 *World population densities*

Population distribution

have been referred to in Section 5.2, but *human and economic* factors are also relevant. Distance from centres of industry and agriculture makes for isolation and *inaccessibility*. Even the presence of natural resources creates only very small local concentrations of population at mining centres.

The explanation for a dense or sparse area of population will always include more than one factor. You cannot answer questions without being specific about regions of the world: you must be aware of the characteristics of different regions and be able to use them in answer to a question.

Questions on the world distribution of population are not generally specific about places. You may be asked to name examples of areas of high or low density, or you may be given a map with some areas marked and asked to select from those. In either case, you would have to give reasons for the population density. This means that you need to have a good general idea about the physical and human geography of different regions of the world.

Distribution by continent

Questions based on continents (or major regions) are usually linked to the distribution of developed and less developed countries, or rich and poor countries (see Unit 14). As well as differences of poverty and wealth, the numbers in each continent differ greatly (as do the growth rates). Always remember that comparisons of continents as a whole hide the variations within each one.

Regional- and national-scale population distribution

What you need to know depends on your syllabus. If particular regions or countries of the world are specified, you may be expected to know particular facts about their population patterns. In any case, you should learn an example from the specified areas. If no particular areas are listed, make sure that you study the population distribution of one or two countries as examples. Questions will ask you to choose and name your own example.

It is useful to learn a set of possible factors which you can use as a checklist. Not all of them would be significant for every example or at every scale of study:

Physical factors	*Human factors*
relief	location of major cities
climate	systems of farming
soils	development and location of industry
natural vegetation	internal and international migration
resources	historical factors
	levels of poverty and wealth
	government policies

Population distribution

QUESTION 7.1 For any country you have studied, describe and give reasons for the distribution of population. (Do not forget to *name* the country.)

QUESTION 7.2 Study the maps below showing the population distribution and other geographical features of Tunisia.
(a) Describe the distribution of population.
(b) How is the distribution of population related to the pattern of rainfall?
(c) Suggest possible explanations for the concentration of population in the north-east of the country.

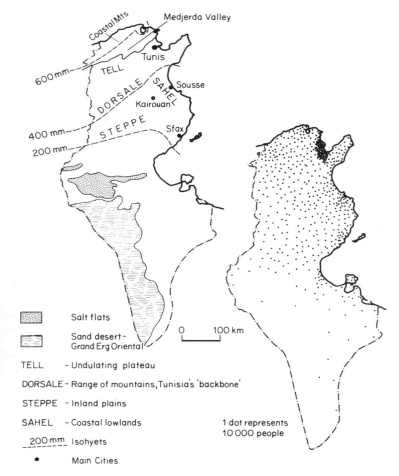

Fig. 7.2 *Tunisia*

87 *Population distribution*

Local-scale population distribution

This scale of study usually concerns the population distribution within towns and cities. Section 9.3 develops other aspects of population patterns besides sheer numbers of people. In this section, the distribution of people is the main concern.

The numbers of people in particular areas of a city are not always helpful, because the size of area varies. Population densities are therefore the only way of making comparisons. Population densities within cities are usually expressed as numbers per hectare (e.g. the population density of Inner London is 77·9 persons per hectare and that of Outer London is 33·5 persons per hectare). In some cases, however, you may find that maps show population densities as numbers per square kilometre (as in Fig. 7.3).

QUESTION 7.3 Study Fig. 7.3
(a) In which part of Bombay are population densities greatest?
(b) Describe the pattern of population densities within that area.
(c) Draw a simple line graph to show how population density changes along line X, Y Z. Write a brief comment on the features shown by your graph.

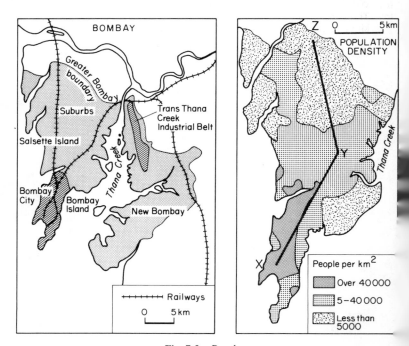

Fig. 7.3 *Bombay*

Population distribution 88

It is also possible to show population densities as a graph drawn along a transect line. Fig. 7.4 shows the typical population density pattern in and around a British city in the nineteenth century.

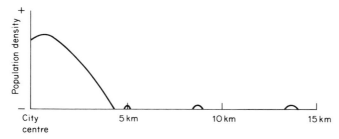

Fig. 7.4 *Population density in and around a typical nineteenth-century British city.*

QUESTION 7.4 Study Fig. 7.4.
(a) Approximately how far does the city extend from its centre?
(b) Where are population densities greatest?
(c) There are small pockets of population shown at about 5, 8 and 13 kilometres from the city centre. What do they represent?

POPULATION
7.2 Size, growth and structure of population

Size
The total population of the world has grown rapidly in the last two centuries. The rapid growth this century has been called the 'population explosion'.

You must be able to *describe changes*. These are usually shown as graphs or sets of statistics. Information about regions of the world may be combined with total world figures, as in Fig. 7.5.

QUESTION 7.5 Study the graph showing the changes in total and regional world population.
(a) Describe the changes in total world population: (4 marks)
(b) Compare the changes in and relative importance of the populations of Asia and Europe over time: (10 marks)

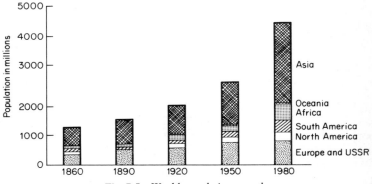

Fig. 7.5 *World population growth*

Part (*a*) is straightforward. Note the number of marks allocated, which gives you an idea of how much time to spend and how many points to make. Use approximate figures derived from the graph to illustrate your answer. Part (*b*) is much more demanding, is therefore worth more marks and should take correspondingly more time.

Growth
Explanation of population growth involves study of vital rates (*birth rates, death rates, growth rates, infant mortality rates, fertility rates*). Except for growth rates, which are given as percentages, these are all stated per thousand (e.g. a birth rate of 44 per thousand or a death rate of 11 per thousand).

The rate of growth of a population depends on the relationship between birth rates and death rates, although migration may also be significant

Size, growth and structure of population 90

in particular countries, regions or districts. The natural growth rate is calculated as follows: birth rate − death rate = natural increase.

The rate at which a country's population is increasing depends on the difference between birth and death rates. Demographers (people who study population) have devised a *model* called the **demographic transition model** to show how the changes in birth and death rates are related to population growth over a period of time. This model shows a pattern of four phases, or stages:

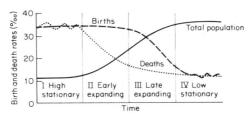

Fig. 7.6 *Demographic transition model*

In *Stage 1*, birth rates and death rates are high and growth slow.
In *Stage 2*, death rates begin to fall but the birth rate remains high. As a result, the natural increase becomes greater and year by year the population grows more and more quickly.
In *Stage 3*, the death rate continues to fall, but more and more slowly; birth rates also fall, slowly at first and more quickly as time goes on. The population continues to grow, but the rate of growth falls.
In *Stage 4*, the birth rate and death rate stabilise at low levels, so that there is little or no growth.

The time scale varies enormously. In Britain, it took over 200 years to go from Stage 1 to Stage 4. Many less developed countries show signs of making this transition much more quickly (e.g. Mauritius has gone from Stage 1 to Stage 3 in 30 years).

During Stage 2, the population begins increaing in size very rapidly and this increase in numbers continues in Stage 3, even though the percentage increase falls. Remember that in Stage 2, the original population is relatively small, but by Stage 3 the country's population is very much larger. Even a small percentage increase then means a large number of people.

The reasons for these changes taking place are not entirely clear, but the following points seem to be true:

In Stage 1, birth rates are necessarily high because death rates are high, due to disease and perhaps poor nutrition.
In Stage 2, improvements in medicine and the prevention of diseases, especially amongst babies and young children, cause the drop in death rates.

Size, growth and structure of population

In Stage 3, birth rates fall, probably because people see no need to have large families: they take steps to reduce birth, because most babies are likely to survive.

In Stage 4, birth rates tend to fluctuate according to social, political and economic circumstances. Death rates however are uniformly low, with the elimination of most deaths in younger age groups.

Structure

The *structure of population* refers to the numbers of males and females in every age group, and it is much affected by changes in the rate of growth of the population as a whole. Structure can be presented simply as the percentage of the population in particular age groups, (e.g. 45 per cent under the age of 15). Usually, however, population structure is shown by a special form of graph called a *population pyramid*.

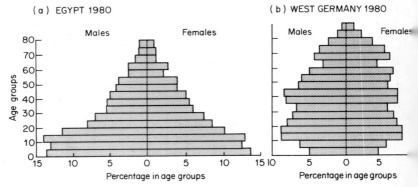

Fig. 7.7 *Population pyramids*

If you compare the two graphs in Fig. 7.7 you will see that A has a very wide base, indicating a large proportion of young people and children, and suggesting a high growth rate. The higher part of the pyramid narrows rapidly, showing fewer people in older age groups. These are the ones who survived the high death rates of earlier years. In B, there is very little difference between the percentage in in any age group until about 60 years of age, indicating a country where growth has been slow or static for a long time and where death rates and birth rates are low. Most deaths are clearly due to old age and degenerative diseases.

These population pyramids are commonly used to indicate problems faced by the countries concerned. When you answer a question on this topic, make it very clear which graph or country you are talking about.

In A, the large young population and relatively small adult population

Size, growth and structure of population 92

mean that a large number of unproductive people have to be supported by a small adult working population. This increases any already existing problems like food production, provision of schools and health facilities. In the future, there will be a greater need for jobs and housing as the young grow up and have their own families.

In B, a relatively large proportion of elderly people must be supported by a small working population. The retired population creates problems of need for suitable accommodation and the increasing need for care as the elderly become more aged. Individual problems include isolation of women; as on average women live longer than men, far more of them live alone. The relatively small population below working age only ensures the continuation of these problems in the future.

Population pyramids can be used to show the population structure of any place: a large city, small town or even a village. They give useful indications of changes taking place. Fig. 7.8 shows the population structure of Hemel Hempstead in 1961. This is a new town (see Section 9.4). During the 1950s the population grew by **migration**, as people, particularly young married couples, moved from London into the rapidly growing town. The structure shows the changes and their effects. The very small top of the diagram represents the elderly people of the 'old' town. The bulge in the 30- to 45-year-old group represents the bulk of the people moving in from London. The narrow section from 15 to 30 represents the children of the 'old' Hemel Hempstead. The enormous base shows the children of the newcomers.

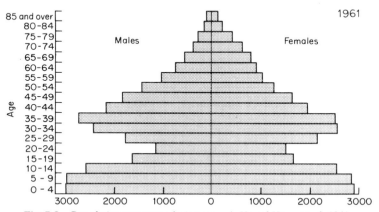

Fig. 7.8 *Population structure of a 'new town': Hemel Hempstead, 1961*

Graphs like this are frequently used in questions about migration, whether into or out of an area, because of the great effect it has on population structure.

POPULATION
7.3 Population movements

Migration

Throughout history, people have been on the move. In recent centuries, the biggest *international* migrations have involved:

1   Europeans colonising North America, South America, South Africa, Australia and New Zealand in enormous numbers.
2   Slaves taken from West Africa to North America, the Caribbean and South America.
3   Chinese moving to South-East Asia and USA.
4   Indians moving to East and South Africa.

In recent decades, international migrations have included:

1   West Indies to UK and, more recently, to the USA and Canada.
2   Southern Europe to Northern Europe.
3   North Africa to Europe.
4   Indian sub-continent to the UK and Middle East.
5   Eastern Europe to Western Europe in the immediate post-war years.

Even on an international scale, not all migrations are permanent. Movement between nearby countries may be to seek employment for only a few years. The reasons for migration at whatever scale can be summarised under two headings: *push factors* and *pull factors* (see the table below).

'Push' and 'pull' factors

Push factors	Pull factors

Escape from:
    poverty
    political persecution
    religious persecution
    environmental disaster
Lack of land
No employment prospects
	Freedom from persecution
	Opportunities for higher living standards
	Better employment prospects
	Land available free from population pressure

Within countries, migration is important. In large countries like the USA, internal migration matches in scale the international migration of Europe. Even within quite small areas, migration causes considerable changes in the distribution of population. Amongst the different types of movement are:

1   Rural to urban migration. This is often spurred on by poverty in rural areas and opportunities (real or perceived) in urban areas, especially large cities. Where out-migration is greater than the natural increase, such areas decline in population. This is called *rural depopulation*.

2 Movements to more prosperous regions. In the USA, there has been a shift of population from the north-east (areas of old declining industry) to the south and west (areas of rapid industrial growth). In the UK, there has been a persistent 'drift to the south-east'.
3 Movements from large cities, especially from inner areas, to the suburbs of the cities and to smaller towns.

There are, of course, *seasonal* movements of people (e.g. the movements of nomadic peoples with their herds of animals).

Urbanisation

Section 9.1 deals fully with this topic, but aspects of it are often included in questions on population. Questions based on a comparison of countries and their population include statistics showing the levels of urbanisation (i.e. the percentage of the population living in urban areas). Generally speaking, developed countries have higher levels of urbanisation than less developed countries. However, less developed countries are becoming more urbanised at a very rapid rate. The table below illustrates this point.

Percentage living in urban areas

	1960	1980
Canada	70	76
Zaire	14	30

Where you are given actual numbers living in cities over a certain size you can estimate the level of urbanisation, provided you are also given the total population.

Questions about population distribution tend to be based on maps, and questions on other aspects of population on graphs or tables of statistics. Depending on the exact wording of the question, you will probably have to do all or some of the following: describe what is shown; explain or give reasons for what you have described; identify problems associated with the points you have made; predict future developments; refer to examples of your own.

SETTLEMENT
8.1 Types of settlement

8

Settlements are places where people live. They can be temporary (as with nomadic camps), or they can be permanent. Permanent settlements can be classified in different ways:

1 *Size* – based on population, or area of land covered, or the range of service present.
2 *Functions of towns* – based on the most important or most obvious function. Types of town include: industrial towns, university towns, county towns, military towns, ecclesiastical towns, mining towns, dormitory towns.
3 *Location features* (where places are labelled, e.g. route centres, harbour sites or bridge-point locations).
4 *Rural and urban* (a very broad division between towns and cities on the one hand, and villages and hamlets on the other).
5 *Nucleated and dispersed* (this describes the patterns of rural settlement, and distinguishes between those settlements consisting of a group of dwellings and isolated farmsteads).

The names used for different settlements do not all have fixed meanings. The term 'city' can mean anything from a place with only a few hundred people in some countries, to a place with a population of millions. Whatever terms you use, it is a good idea to define your meanings. Definitions are usually based on the size and function of a place. Here is a commonly-used set of names:

1 Individual: separated farm or house.
2 Hamlet: small cluster of houses and farms with very few, if any, services or functions besides residential and agricultural.
3 Village: rather larger and covering a wide range in size of population. Always has a range of basic services.
4 Town: the first urban settlement type, larger with a wide range of service functions and some industry.
5 City: larger, with a greater range and number of services and more industry.
6 Metropolis/capital: the region's largest city, with the greatest range and number of functions.

Towns and cities cover a wide range and some people would include 'regional centre' between nos. 5 and 6. A regional centre is a very large city and, as the term suggests, the major settlement in a particular region of the country.

QUESTION **8.1** Study the Ordnance Survey map extract. Identify and name: farm, a hamlet, a village and a town. Describe each under the headings *siz* and *functions*, using map evidence only. (As a guide, look at examples and as yourself, 'What functions or services are there in each place?')

8

SETTLEMENT
8.2 Site and situation

The *site* of a settlement is the land on which it is built. In the case of an old settlement which has grown, the term normally refers to the original site.

Situation refers to a settlement's general setting, or its location in relation to major features like hills, valleys and river estuaries, routeways or other major settlements.

Site

Sites reflect the needs of a settlement at the time of establishment. With rural settlement, two factors related to water have always been important: (*a*) a site should have drinking water available, (*b*) a site should not be in danger of flooding. In chalk and limestone areas of the country, it is usual to find settlements at spring lines at the foot of scarp slopes (see Fig. 8.3), and on the dip slopes settlements are found along valley bottoms. In low-lying and wetter areas of the country, settlements avoid areas likely to flood and are found on raised, better-drained land.

Questions to do with settlement sites are commonly part of map-reading sections in examinations.

QUESTION 8.2 Study the Ordnance Survey map extract and comment on the sites of settlements along the Severn valley. (Not the general absence of settlement on the flood plain.)

Site factors for towns involve other points: defence was often an important factor in medieval times, as was an easy river crossing point, or a safe harbour for a fishing town or port.

Situation

Situation is important, as towns must be accessible from the surrounding area, if they are market or government centres. They must have access to resources or markets if they are industrial centres, and transport and communications if they are ports.

QUESTION 8.3 Study the sketch map below and write a brief description of the situations of towns A and B.

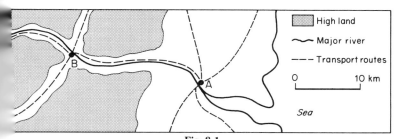

Fig. 8.1

Types of settlement

Major cities

Major cities have become important for many reasons but location is the most important factor. Originally, the site may have been vital in ensuring the city's initial success and continuance. For long-term growth, the situation within the immediate region and the country is most important.

All major cities are centres of communication, because all the functions of a major city depend, eventually, on good communications. Large cities have many functions, and the larger the city the greater the area for which it provides these functions. London, for example, has the following:

government	industry
shopping	residential areas
tourism	entertainment
port – sea and air	land transport

These functions are matched on a smaller scale by other, less important cities. Remember that the capital city with government functions is not always the major city. In a number of countries, a capital city has been established that is separate from other cities. Washington in the USA, Canberra in Australia, Islamabad in Pakistan and Brasilia in Brazil are all such examples.

Questions to do with site, situation and functions are often part of map-reading questions. Otherwise they are likely to be short or multiple-choice questions.

QUESTION 8.4 The map below shows an area in which a number of settlements have developed. *One* has grown into a major city. Choose the settlement most likely to have become a major city, and choose the reason which is most likely to explain its importance.

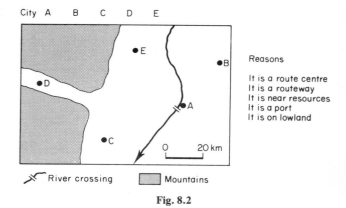

Reasons

It is a route centre
It is a routeway
It is near resources
It is a port
It is on lowland

Fig. 8.2

8

SETTLEMENT
8.3 Patterns and hierarchies of settlement

Patterns of settlement vary according to physical and human (including historical) factors. A simple division is into *nucleated* and *dispersed* patterns. In a nucleated pattern, all the settlements consist of clusters of dwellings. In a dispersed pattern, individual farms and houses are spread separately throughout the area. In most of Britain, the settlement pattern is a combination of nucleated and dispersed settlements. Two other ways of describing the pattern are in terms of *density* and *spacing*. The greater the density of settlement, the more closely peopled is the area and the smaller the average distance between settlements.

Physical factors are usually related to relief, drainage, geology and soils, or a combination of these. In mountain areas, villages are arranged along the valleys; in areas with richer soils and generally more favourable conditions, settlement is denser than in areas of poor soils and harsh climate; in areas where water supply was a problem in the past, settlements are found along spring lines.

There are many historical factors. A generally dispersed pattern may reflect an ancient landholding system or an ancient farming system; a highly nucleated pattern may be the result of the need for people to group together for defence, or to work the land together in a common farming system; dispersed settlement on recently reclaimed land is a result of the ability to travel greater distances to towns and villages than in the past.

QUESTION **8.5** Describe, and suggest possible reasons for, the settlement patterns shown on the maps below.

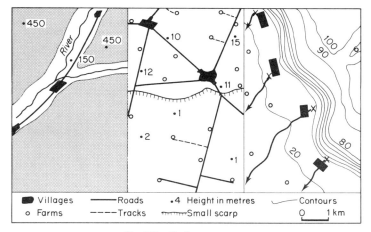

Fig. 8.3 *Settlement patterns*

99 Patterns and hierarches of settlement

Settlements are arranged in *hierarchies* according to their size or their range of functions. Functions refer to services: there are, in any area, a large number of villages with limited service functions, few towns with a wide range of functions and one large town with a very wide range of functions.

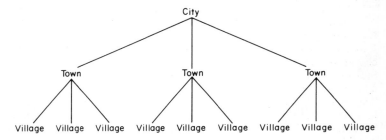

Fig. 8.4 *Settlement hierarchy*

The set of ideas to do with the pattern and organisation of settlements as service centres is called **Central Place Theory**. A central place is a settlement which provides services for the surrounding area. If Fig. 8.4 were presented as a map, the theoretical settlement pattern would be like that in Fig. 8.5. Each settlement has a market area which is hexagonal in shape. Each village serves a small surrounding area. Towns serve one-third of each of their neighbouring villages. The city serves one-third of each small town and the appropriate surrounding area.

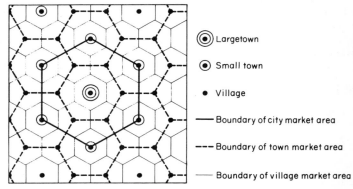

Fig. 8.5 *Settlement hierarchy: theoretical pattern based on hexagonal market area*

Patterns and hierarches of settlement

The key ideas of this topic are: order of services, range of services, sphere of influence and threshold.

Order of services Services or goods needed frequently and for which people are not prepared to travel any distance are called *low-order* or *convenience goods*. Those that are needed less frequently and for which people will travel a greater distance and will want to make comparisons are called *high-order* or *shopping* or *comparison goods and services* (e.g. bread or toothpaste would be low-order while furniture or clothes would be high-order goods).

Range of a service This refers to the distance people will travel for a certain service.

Sphere of influence This is the area from which people travel to buy goods and services in a particular settlement.

Threshold This is the number of potential customers needed for a particular business to be profitable. Goods and services needed frequently have a lower threshold and those needed less frequently have a higher threshold. Therefore, a village and its sphere of influence might have enough people to support a butcher's shop but not enough to support a furniture shop.

Note the difference between a 'range of services' meaning how many types of service, and the 'range of *a* service' meaning the distance people will travel for a particular service.

Questions on this topic use maps showing settlements in an area, and give information like the road network and perhaps the population (as a guide to the range of functions). They also use two other ways of presenting information: scalograms and desire-line maps (Fig. 8.6). Scalograms show you the functions that each settlement has, and you can use them to distinguish the high-order settlements from the low-order ones. A 'desire' line or 'trip' line joins the home of a local person to the place he or she visits for a particular service. Desire-line maps show you the sphere of influence for particular service. A comparison of the distances travelled or sizes of the sphere of influence also separates higher-order from lower-order settlements.

101 Patterns and hierarches of settlement

QUESTION 8.6 Study Fig. 8.6.
(a) Which place shown on the scalogram represents the highest-order settlement? Give a reason for your answer.
(b) Which of the desire-line maps shows a low-order function? Give a reasons for your answer.

Function Settlement	1	2	3	4	5
A	5	6	3	4	3
B	3	2	1	2	–
C	1	1	–	–	–
D	1	2	–	–	–
E	1	1	–	–	–
F	1	1	–	–	–

(a)

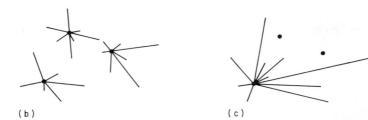

(b)   (c)

Fig. 8.6 *(a) Scalogram of matrix (b) Desire-line map for visits to grocer (c) Desire-line map for visits to shoe shop*

Patterns and hierarches of settlement 102

QUESTION 8.7 Study the map below, showing the hierarchy of settlement in an area.
(a) Which places are the highest-order settlements?
(b) On the map, show the probable spheres of influence or market areas of those places.
(c) Name a higher-order settlement you have studied and describe its range of functions and sphere of influence.

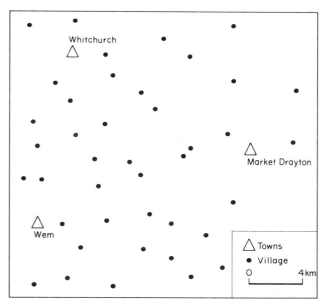

Fig. 8.7

SETTLEMENT
8.4 Changes: decline and growth

Settlements can change quite quickly and the changes can come about for a variety of reasons. Changes in large cities are dealt with in Unit 9. Here, we shall look at changes affecting villages and small towns.

Decline
Generally, it is villages that decline most markedly and this is usually a result of rural depopulation, or the closure of the main local sources of employment. Villages in Britain have lost population because:

1 Farming jobs have disappeared and farms themselves have been amalgamated.
2 Public transport in rural areas is poor or non-existent.
3 Alternative work is only to be found in the towns.
4 The isolation of rural life is a problem, especially for families.

These factors have combined to see many villages, especially the more isolated ones, lose much of their population, and particularly younger people. The effects of population loss have been:

1 Loss of services, like village shops and schools.
2 Further loss of public transport.
3 Derelict houses and farm buildings, giving the village a run-down appearance.

Villages based on a single source of employment, like coal-mining or quarrying, have suffered the same sort of problem. In such cases the change was instant, not gradual. Sometimes, local planning authorities have tried to minimise such effects by concentrating people from villages that are becoming depopulated into one main village, where services can be maintained.

Growth
The villages and small towns that have experienced growth tend to be:

1 Accessible to large towns and cities and providing houses for commuters in converted or renovated older buildings, or in new housing estates;
2 in areas where some new development acts as a stimulus (e.g. the development of tourism)

Places like this change by:

1 Growth of population.
2 Increase in area and number of buildings.
3 Increase in number of services in the case of towns, but not normally in the case of villages (since newcomers tend to travel by car to the towns for shopping).

Changes: decline and growth 104

4 Change in the social structure: many of the newcomers are likely to have relatively well-paid jobs, or are perhaps retired professional people rather than the farmworkers they replaced.

Questions on this topic tend to be quite straightforward but, as always, the information you are given can be presented in many different ways, e.g.:

maps showing an individual village before and after changes
a newspaper cutting describing some development
tables of statistics or graphs of population or employment
maps of an area and local transport timetables
photographs before and after a change
advertisements for individual houses or estate developments

QUESTION 8.8 Study the map showing the growth of a village (Fig. 8.8).
(a) Describe the changes in the village.
(b) Give reasons for the village's growth and the other changes that have occurred.

(Note that the wording here does not give detailed guidance about what to include in your answer. In part (*a*), *all* changes should be included and in part (*b*), give every reason that is relevant.)

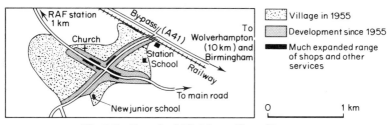

Fig. 8.8 *Growth of Albrighton, Shropshire*

QUESTION 8.9
(a) (i) Study the diagram (Fig. 8.9) and put the labels in the appropriate boxes.
(ii) Explain the changes shown in the diagram.
(b) Describe and explain the changes in a settlement you have studied.

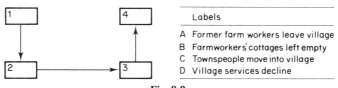

Fig. 8.9

URBANISATION
9.1 Urban trends

The world's urban population has increased enormously during the last century. The growth of cities has been linked to industrialisation and trade. The growth of the populations of cities and migration to cities from rural areas have both been important. In Section 7.3, urbanisation was defined and the general difference between developed and less developed countries was shown. However, the situation is changing rapidly and less developed countries are becoming urbanised at a very rapid rate.

Questions to do with this general topic are usually concerned with:

1 Comparisons between rates of urbanisation in different countries.
2 Distributions of large cities on a world scale, using tables or maps showing 'million' or 'multi-million' cities.
3 Comparisons of the numbers and patterns of large cities over time (e.g. compare the number and distribution of 'million' cities in 1935 and 1985).
4 Locations of the world's fastest-growing cities, which are all in less developed countries, particularly in Latin America and Asia.

Such questions are largely *data-response* questions: they test your ability to extract information from various types of material, and they also test your knowledge and understanding of the topic in general. Sometimes, part of the question goes on to ask for details of examples you have studied.

The largest cities have often grown into **conurbations**. These are continuous built-up areas formed *either* as small towns and villages are engulfed by a growing city, *or* by the growth of a group of neighbouring towns and cities all of which merge with one another. Examples of the former are London, Paris and Tokyo. Examples of the latter are the West Midlands, the Ruhr (West Germany) and Randstad (Netherlands). Questions to do with conurbations in developed countries, especially Britain, are usually concerned with planning (9.5) or with recent changes in population densities (9.3).

The other aspect of urbanisation which is particularly important to Britain and other developed countries involves movements of people in recent years. The big cities and their suburbs have lost population. Small towns near the big cities and conurbations have grown, as have towns in more rural areas. The question on page 107 illustrates the changes, which are linked to changes in the location of industry as well as people's preferences for places to live.

Urban trends 106

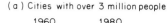

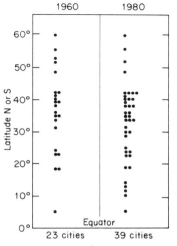

1 dot represents 1 city with over
3 million people

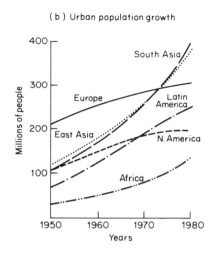

(c) Population of Greater London (in millions)

	1971	1981	% change
Inner London	2.9	2.4	−18
Outer London	4.4	4.2	−5
Greater London	7.4	6.6	−10

(d) Population of Madras (in millions)

1951	1.4
1971	2.5
1981	4.3

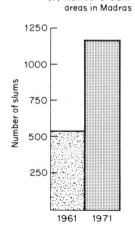

Fig. 9.1

Urban Trends

QUESTION 9.1 Study the information in Figure 9.1 which is concerned with the world's very large cities.

(a)   (i) How did the number and distribution of major world cities with over 3 million people change between 1960 and 1980?

    (ii) How does the graph showing urban population growth by continent in Fig. 9.1 (b) help to explain your answer to part (a)(i)?

(b)   (i) Comment on the changes in population of London and Madras as shown in Fig. 9.1 (c) and (d).

    (ii) In what ways do the changes in population in London and Madras confirm your answer to part (a)(i)?

(c)   (i) Fig. 9.1 (e) shows the growth of slums in Madras. Give two of the names by which such areas are known.

    (ii) The growth of the slums is linked to migration into Madras. From where are the migrants most likely to have come?

    (iii) Suggest reasons why large cities like Madras attract so many migrants.

    (iv) What problems are caused for the city itself and for the migrants by the rapid growth of the city by migration?

Note that parts (a) and (b) depend on your ability to interpret the information given, but part (c) calls for you to use your knowledge of the topic. (See sections 7.3 and 9.5.)

9

URBANISATION
9.2 Patterns within cities

Patterns within cities are concerned with:

1 *Age zones*, with each consecutive time period showing the growth of the city, and with differences in road patterns and housing types between each age zone.
2 *Functional zones*, with the arrangement of the different land uses and the relationships between them being the centre of attention.

A model of growth
In general, cities grow in concentric rings or age zones, that is, with each new stage of growth forming a ring around the earlier one. A simple **model** of city growth is shown in Fig. 9.2(*a*). Of course, real cities do not grow in perfect rings like this but as Fig. 9.2 (*b*) shows, the general idea is applicable to London and also to the majority of other cities.

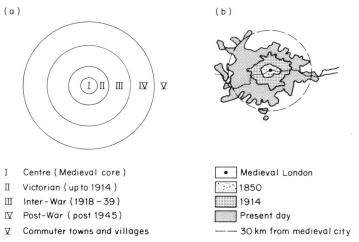

I	Centre (Medieval core)
II	Victorian (up to 1914)
III	Inter-War (1918–39)
IV	Post-War (post 1945)
V	Commuter towns and villages

•	Medieval London
░	1850
▓	1914
■	Present day
———	30 km from medieval city

Fig. 9.2 *Urban growth*

The following points apply to most British cities:

1 The present city centre roughly represents the entire original city, which may have been walled in medieval times.
2 The first big phase of growth came in Victorian times.
3 The second phase of spreading came between the World Wars I and II. Suburbs expanded at low densities, aided by the development of cheap public transport in the form of trams, trolley-buses, electric trains and buses.

4 Since World War II, cities have grown still further, spreading outwards and filling in spaces left in earlier phases of growth. Much of this growth was necessary to accommodate people displaced from the Victorian zone.

5 The growth of commuter towns and villages in a ring around the city has also been a common feature of the post-1945 period. Increased car-ownership has been particularly important in their development.

6 In many cities, the Victorian areas have been *redeveloped*: they have been cleared and the rows of terraced houses replaced by high-rise flats or, as has happened in many areas, left as derelict land. This is also known as *urban renewal*.

Questions are often concerned with planning issues (see 9.5), or with straightforward identification of the features of different zones, especially as they are shown on maps. These may be sketch maps (as in Fig. 9.3), or they may be small extracts from large-scale Ordnance Survey maps or plans. In particular, these questions emphasise the patterns of roads in different phases, the amounts of open space, the density of building and the mixing or separation of industry and housing. A typical short question would ask you to match the maps in Fig. 9.3 to locations on a map of the city or to named periods of growth. Note that you can apply these ideas even to a small town like Welshpool (see OS map extract). If a town, especially a sizeable one, is shown on a map extract, it is common to ask a question of this sort.

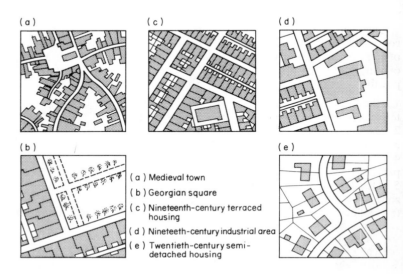

(a) Medieval town
(b) Georgian square
(c) Nineteenth-century terraced housing
(d) Nineteenth-century industrial area
(e) Twentieth-century semi-detached housing

Fig. 9.3 *Patterns of building*

A model of functional zones

There are several models of functional zones of cities and they all incorporate ideas about the distribution of different types of land use *and* the stages of growth. The most useful one for British cities is shown in Fig. 9.4. Of course, real cities do not match the model perfectly, but its features are common to the majority.

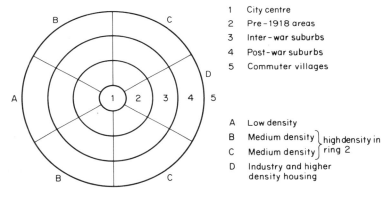

1 City centre
2 Pre-1918 areas
3 Inter-war suburbs
4 Post-war suburbs
5 Commuter villages

A Low density
B Medium density ⎫ high density in
C Medium density ⎭ ring 2
D Industry and higher density housing

Fig. 9.4 *City zones model*

Questions about the different functional zones of cities are sometimes concerned with just one function, like industry, or with the overall pattern of all functions. There are many different ways of presenting the material on which questions are based, but maps are the most common. The types of possible questions include:

1 Short questions/multiple-choice questions asking for definitions of terms or identification of features or zones.
2 Full questions on the growth and layout of a city.
3 Full questions on one type of functional zone.
4 A problem-solving question, in which you choose a location for a particular activity and justify your choice.
5 Questions incorporating basic ideas about urban zones (i.e. functional zones) with planning problems and issues, and urban problems generally. (See following sections.)

111 *Patterns within cities*

Urban functional zones

Services
The *central business district* (*CBD*) is the city centre, and contains the single most important concentration of shops, business and government offices and entertainment facilities. In all but the smallest towns you can identify specialised zones (e.g. office zones and shopping zones) as particular functions tend to cluster together. The CBD is an accessible location and land values are high as a result of competition for space. This is reflected in the more intensive use of land, with taller buildings than elsewhere.

Shopping centres are arranged in a hierarchy within the town or city, with the CBD as the top level. Local shopping centres with only a few shops are the bottom level. Depending on the size of the city there will be several levels of centre in the hierarchy (e.g. there may be local, district and major shopping centres besides the CBD itself). Note that shopping centres can also be called *business centres*, since the larger ones have financial and general office functions as well as shops.

Questions which ask for examples are ideally answered from your own experience. Make a simple map of the shopping/business centres in your town to show the different levels of importance. You can use the scalogram technique shown in Section 8.3 in your answer.

Industry
The pattern of industry reflects the time when industrial areas were established:

1 Old industry in the Victorian zone tended to be mixed up with housing and close to railways and canals.
2 In the inter-war period, industry established itself in and on the edge of the suburbs on industrial estates adjacent to arterial roads, since road transport was becoming more important.
3 Modern industry in cities is more likely to be located on industrial estates – still separate from housing areas, but on the edge of the built-up area, and again close to roads.
4 Heavier industries tend to be situated along railway lines or along rivers where they are navigable.

Patterns within cities 112

Residential areas

The Victorian zone of a city has mixed industry and housing. The housing was in terraces except in the better-off areas, where larger houses have now been converted into offices in many cases. Much of this zone has been redeveloped, so as well as the older terraced housing (which was very high density), modern high-rise flats are the characteristic feature.

The inter-war suburbs – whether private or council houses – were built to a much lower density. The road plans were of varied geometrical shapes. Post-war suburbs have a more varied street layout, with a higher density of building than in the inter-war period.

QUESTION **9.2** Which of the following most accurately defines the word *redevelopment*?
A The widening of major roads through a city.
B The development of a shopping centre on the edge of a city.
C The clearing and replacement of old housing and industrial areas in a city.
D The renovation of houses in particular parts of a city.

QUESTION **9.3** Which of the following most accurately describes the part of a city which contains a great variety of shops and offices clustered together in a relatively small area, and attracting people from all over the city and beyond its boundaries?
A District shopping centre.
B Out-of-town shopping centre.
C Business park.
D Central business district.

URBANISATION
9.3 People in cities   9

Questions are concerned either with the distribution of population in cities, or with differences in poverty and wealth within cities. Such questions would be part of broader questions covering other aspects of cities.

The following general statements can be made:

Population
1 The city centre is, in population terms, empty.
2 Patterns of density show that inner-city areas are more densely population than outer areas (suburbs).
3 The outer areas generally have more open space per person than inner areas.
4 As a result of redevelopment (renewal) of inner-city areas and the rehousing of *overspill* population on the edges of cities and beyond, the inner-city areas have been losing population rapidly.
5 In the last decade, the outer suburbs of large cities have also been losing population.
6 Minority groups tend to be concentrated in inner-city areas.

Fig. 9.5 summarises population changes in model form.

Poverty and wealth
1 There is a strong relationship between social status/levels of income and quality of environment.
2 Inner-city areas have greater unemployment, poorer housing, lower car-ownership and poorer health levels.
3 Some large outer-city council estates have the same features as inner-city areas, with the added disadvantage of isolation.

Questions on this topic are generally factual: you are asked to interpret data presented in maps, tables, graphs or photographs. Table 9.1 shows typical census statistics. These are for local government districts, but the same figures could be based on smaller areas such as wards or parishes, and even enumeration districts. These are very small areas, which may include only one or two hundred households. You are often asked to make general statements from the data and to identify relationships (e.g. between location and patterns of unemployment). The explanations for such relationships are complex. However, here are some examples of questions and answers at the level you can expect:

(*a*) Why has unemployment risen in inner-city areas? (Most industry has closed or relocated elsewhere.)
(*b*) Why are there generally poorer health levels in inner-city areas? (Housing is old, in poor repair, with much dampness. Some of the new housing built in the last 20 to 30 years is in a similar state.)

People in cities 114

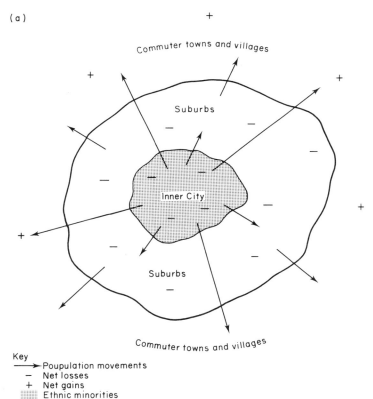

Key
→ Population movements
− Net losses
+ Net gains
▒ Ethnic minorities

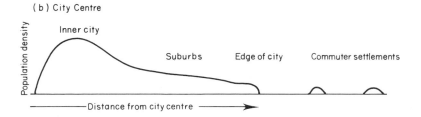

Fig. 9.5 *A model of population change in cities: (a) Map (b) Cross-section*

People in cities

(c) Why has the inner city lost so much population? (Redevelopment removed a great deal of housing and those who could afford to do so have moved elsewhere, where the quality of environment is better.)

In all cases, these brief answers should be illustrated with examples; general statements, like 'quality of the environment' in the last answer, should be defined.

QUESTION 9.4 Study the table below, showing selected characteristics from the 1981 Census for five areas of the country.
(a) Which area has the largest population?
(b) Which area is the most crowded and which is the least crowded?
(c) In which area is there a large percentage of retired people?
(d) Which area has the highest level of male unemployment?
(e) Which characteristic or characteristics might be used to indicate the general level of wealth in an area? On the basis of your chosen characteristic or characteristics, which area seems to be the wealthiest?
(f) In which area or areas do you find the greatest housing problems? Explain how the census characteristics suggest this.
(g) Which areas have experienced the greatest inflow of people from other countries?

Selected census statistics for five areas

	Area				
	Islington (London)	Liverpool	Leicester	Solihull	Eastbourne
Population in thousands	157·5	503·7	276·2	197·9	74·1
Population density per hectare	108·0	45·2	38·2	11·0	17·6
% of pensionable age	17·2	18·5	17·4	13·5	33·5
% born outside the UK	24·8	3·3	19·0	3·8	6·4
% unemployment amongst men	15·8	24·4	16·2	10·9	9·1
% of households with 2 or more cars	5·2	6·5	8·4	25·1	10·8
% households with 1 or more persons per room	26·6	17·0	17·8	10·9	11·4
% households without bath or shower	2·4	3·5	1·9	0·2	1·2
% households without inside WC	1·2	5·2	7·1	20·4	0·8

9

URBANISATION
9.4 Urban problems and planning

Problems include:

pollution
traffic
housing
public open space
dereliction
congestion and overcrowding
unemployment

Planning in and around cities has involved:

traffic
redevelopment of CBDs
redevlopment of inner-city areas
new towns and expanded towns
green belts

The problems overlap, as do the planning issues. However, it is useful for you to consider some of them separately.

Traffic problems
These have arisen largely from an increase in road traffic (private cars and lorries) so that congestion of city roads is common, together with noise and atmospheric pollution. The solutions have been:

1 By-passes and ring-roads to keep through traffic out of the city altogether.
2 Inner ring-roads encircling the CBD to keep through traffic out.
3 Pedestrianising streets in the CBD.
4 Using one-way systems to discourage traffic from entering the CBD.
5 Providing parking spaces outside the CBD.

Only a few areas have positively encouraged public transport as a means of reducing the number of private vehicles.

Redevelopment of Central Business Districts
In many cities, large areas have been cleared and new shopping areas, often roofed-in, heated and air-conditioned, have been built. These places usually have separate access for goods vehicles, and the shopping areas are traffic free. One effect of this development has been the run-down of some other parts of the CBD. Another has been the destruction of many older buildings of character and architectural value. Property development generally has changed the skyline of most large cities in Britain, so that tall office blocks and hotels are common.

Urban problems and planning

Redevelopment of inner-city areas
This has been examined earlier, but it is important to note a change in planning approaches in such areas. High-rise blocks of flats are no longer built. Apart from the structural problems of these buildings, it is now recognised that they cause many social problems. In recent years, there has been more emphasis on the renovation of existing property, together with efforts to increase employment in such areas.

New towns and expanded towns
The redevelopment of inner-city areas meant that the overspill had to be accommodated elsewhere. The aim was to reduce congestion as well as improve housing and living conditions generally. People were encouraged to move to new and expanded towns, and so was industry. Around London, a ring of New Towns was established in order to take relocated people and industry from the city. New Towns were based on small existing settlements, whereas expanded towns were exactly what the term suggests – an expansion of an already sizeable town. New Towns introduced many new ideas in planning, including: pedestrianised shopping centres; neighbourhood housing areas centred on local shopping, social and educational facilities; limitations on traffic within residential areas, the separation of through roads from housing areas; the separation of housing areas from industrial areas. Note that New Towns are common throughout the world, so examples in examination questions could be drawn from widely different areas. However, these towns are almost all designed on readily identifiable principles.

Green Belts
These form part of the overall set of planning policies that produced new towns. They are basically rings of land around major cities and conurbations where there are strict controls on new development. The aim was to stop the sprawl of cities across the countryside. In Britain, Green Belts have encircled cities, but in other countries variations on the same idea have included green 'wedges', with corridors of development between them.

QUESTION 9.5 What problems have been caused in towns by increased road traffic? With reference to examples you have studied, describe the methods which have been adopted in an attempt to solve these problems.

9

URBANISATION
9.5 Cities in less developed countries

The fastest-growing cities in the world are all in less developed countries. They are growing quickly due to a combination of high rural-to-urban migration and high natural rates of increase.

There are major differences between the cities of developed and less developed countries. A common *model* of a city in a less developed country is shown in Fig. 9.6 This general pattern is repeated in many cities. There are differences, however. The largest also have central city slums or shanty town developments on any available site, even between luxury apartment blocks. The variety of historical, cultural and physical backgrounds produce many other differences: there are huge urban agglomerations, like Calcutta and Mexico City; cities based on traditional cores, like Cairo and Tunis; and planted cities, like Lusaka.

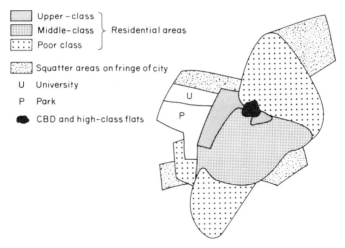

Fig. 9.6 *Model of cities in less developed countries*

All cities have differences which are due to local factors. For example, Mexico City, set in a basin surrounded by higher mountains, has a massive pollution problem. Its site on a drained lake bed also makes it more vulnerable to the effects of earthquake, as the 1985 earthquake showed. Lagos has problems resulting from a site on marshy land at sea level around a lagoon. The common features which override individual differences are rapid growth and high in-migration. Newcomers from rural areas are attracted to large cities, often ignoring small local towns. Migration is a result of the appalling conditions of poverty in the countryside and the

Cities in less developed countries

higher employment prospects in the cities. Work, where it can be found, may be in industry for the more fortunate, but is more likely to be in services. A very high proportion of newcomers work in services. This is usually taken to be a key indicator of an advanced economy. These jobs, however, are largely as street pedlars, gardeners, domestic servants and so on.

The newcomers to the cities need housing. Without land or money, they have to squat on any unused land and build their own homes from whatever materials are available (for nothing, if possible). The squatter areas are often on marshy land or steep slopes, and as shown on the model, on the fringe of the city. There are many different names for the squatter areas, including *bidonvilles*, *favelas* and *barriadas* as well as 'shanty towns'.

Questions on this topic are likely to:

1 link with population topics, like rural depopulation and migration generally;
2 concentrate on the growth of an individual city;
3 compare cities in developed and less developed countries;
4 focus on specific problems of cities in less developed countries.

You could well be asked for details of an example you have studied. Depending on your syllabus, this could be any city or one from a specific area of the world.

QUESTION 9.6   What is a shanty town?

QUESTION 9.7   Study the sketch map below, showing a city in a less developed country (Fig. 9.7).
(a) (i) On a tracing of the map, label with a letter A the area where the richer people of the city are mose likely to live.
(ii) Label with a letter B an area on which a squatter settlement is likely to develop. Give reasons for your choice of locations.
(b) Suggest how the local authority could help to reduce some of the problems of the residents of the squatter area. Refer to examples you have studied.

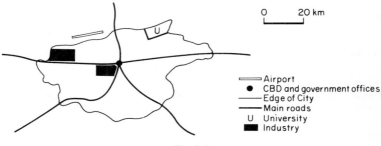

Fig. 9.7

10

TRANSPORT
10.1 Distance, journeys and types of transport

Distance
There are different ways of considering **distance**:

1 *Linear distance* is the ground distance between places, measured in kilometres or other units. It can be for direct, straight-line distance or for actual distance travelled along roads.
2 *Time distance* measures a journey between places by the time it takes to travel.
3 *Cost distance* uses the amount of money involved as the measure.
4 *Perceived distance* is the distance people think or feel there is between places or to a place. This might be a measure of the difficulty of the journey or the level of geographical knowledge of the person concerned.

A common relationship for many types of journey is what is called **distance decay**. This means that there are more short-distance journeys made to a particular place than long-distance ones. Applied to people's journeys to work, this means that more people travel a relatively short distance to work at a place; the numbers fall off rapidly as the distance increases.

The types of transport used depend on their relative advantages. The table below shows the advantages of different modes of transport for goods.

Advantages of different modes of transport

Mode of transport	Advantages and uses
Water (sea or inland waterways)	Carry large amount at low cost; useful for bulky and low value products, where speed is not important
Railways	Moves large quantities cheaply and quickly; useful for bulky materials over medium to long distances but not on a local scale
Roads	Fast and flexible; useful for door to door deliveries of small- to medium-sized products, especially on a local scale
Air	Fast; useful for any traffic where speed is important and for products which are valuable in relation to their weight
Pipelines	Carries liquids where continuous flow is possible; useful where a large volume is needed constantly

QUESTION 10.1 For the transport of people, write down one advantage and one disadvantage of each type of transport.

121   *Distance, journeys and types of transport*

QUESTION **10.2**   What means of transport would be most appropriate for sending a piece of video equipment from Newbury in Berkshire to Los Angeles? Give reasons.

QUESTION **10.3**   What means of transport would be most appropriate for taking 5000 tonnes of alumina from Jamaica to Lynemouth in Northumberland?

Transport nodes

A **node** is any location where transport routes focus. In this section, it could mean a bus station, railway terminus and, in particular, a seaport or airport.

Seaports develop:

1   Where natural features provide a natural harbour, or where there is scope for building a man-made one.
2   Where there is sufficient space for port facilities (e.g. warehousing and goods yards for road and rail).
3   Where there is access to a *hinterland*, which will support the port's activities.

Requirements change with new developments in the size of vessels, especially bulk carriers, and the growth of container traffic. They both concentrate traffic at fewer ports.

Airports require:

1   A large area of generally level land.
2   Access to a large work-force.
3   Good links with centres of population.

Airport development causes controversy because of:

1   Loss of farmland, both to the airport itself and to linked housing, roads and industry.
2   Homes being demolished.
3   Aircraft noise.
4   Extra road traffic.

Roads and motorways

Questions concerning these may be to do with networks generally (10.2), planning issues and policies (10.3), particular features of their routes (often in map-reading questions) or their unique features. This applies particularly to motorways, which have: (*a*) limited access, (*b*) high speeds, (*c*) no cross-roads, (*d*) limited gradients.

10

TRANSPORT
10.2 Routes and networks

Individual *routes* between places generally follow a near straight line unless they are forced to deviate or detour. Note that:

1 Negative deviations or detours avoid obstacles or difficulties.
2 Positive deviations or detours turn towards some attraction.

These ideas are common in map-reading questions about routes followed by roads and railways.

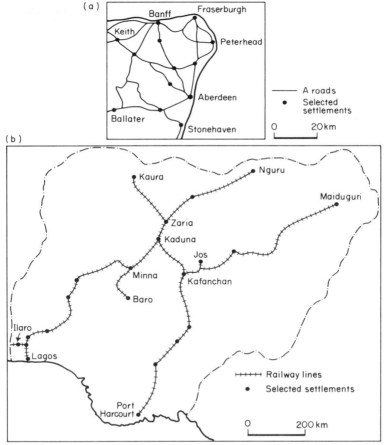

Fig. 10.1 *Transport network: (a) Road network in north-east Scotland (b) Railway network in Nigeria*

Routes and networks

Networks of communication develop as individual routes join each other. Two basic types of network are *branching* networks and *circuit* networks. The more complete a network, the more circuits there are. Such networks have a high level of **connectivity** – i.e. they are better connected.

Questions sometimes compare networks in developed and less developed countries. Networks in the latter are often less complete and this gives rise to some problems, especially in transporting goods.

Accessibility compares how easy it is to reach particular places. The idea of accessibility is very useful for discussing the location of towns and the best locations for various kinds of economic activity.

Measuring networks and routes These measures are useful for making accurate statements:

connectivity index = $\frac{\text{no. of links}}{\text{no. of nodes}}$ (less than 1 is a branching network, 1 has one circuit, higher numbers mean higher levels of connectivity)

centrality index = no. of roads and railways meeting at a place; if appropriate, give extra weighting to main roads

detour index = $\frac{\text{distance along route}}{\text{straight-line distance}} \times 100$ (the higher the number above 100, the greater the detour)

Accessibility is measured by completing a matrix of distance between each place in an area. Totalling these distances gives a measure of comparison (*accessibility index*), and the place with the lowest total is the most accessible (Fig. 10.2).

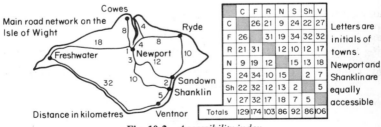

Fig. 10.2 *Accessibility index*

QUESTION 10.4 Study Fig. 10.1, showing two contrasting transport networks:
(a) Calculate the connectivity index for the railway network of Nigeria.
(b) Compare the centrality of Aberdeen, Peterhead and Ballater in north-east Scotland.
(c) Calculate the detour index for the railway journeys between: (i) Nguru and Kaura, (ii) Port Harcourt and Lagos, (iii) Zaria and Minna.

10

TRANSPORT
10.3 Transport planning and policies

Developments in transport have far-reaching effects. Here are some changes. Some of them have been taking place gradually over 50 years or more, whilst others are very recent:

1 Closures of railways in the UK after 1962 largely affected branch lines and rural areas.
2 Improvements to railways, like electrification and high-speed trains, shorten journey times between cities.
3 Building of motorways shortens journey times.
4 Motorways made roads compete with railways in journey times.
5 Increased numbers of cars and lorries have created congestion and other environmental problems.
6 Increased use of cars has helped bring about reductions in public transport, especially in rural areas.
7 Poor public transport affects most those people without their own means of transport.
8 Development of airports creates many new jobs.
9 Airports create major environmental problems, especially noise.
10 Development of containers has led to the concentration of traffic in fewer ports and a loss of jobs.

This list could probably be added to indefinitely. See if you can add a few more items yourself (some have been mentioned earlier in the book). The changes are so many and so varied that most questions give you information from which you can work out the effects. However, you may be asked to describe an example you have studied. Questions very often expect you to weigh up the arguments for and against a particular development.

QUESTION 10.5 Fig. 10.3 shows some of the changes brought about by increased car-ownership and travel.
(a) Some groups of people have 'limited mobility'.
   (i) Name two groups of people whose mobility is limited in some way.
   (ii) Explain why the mobility of these groups is limited.
(b) Choose one of the ill-effects indicated on the diagram and suggest some measure that could be introduced to remedy it.

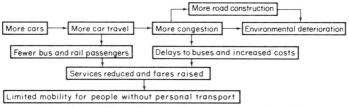

Fig. 10.3 *Effects of increased car-ownership and travel*

INDUSTRY
11.1 Types of industry

Industry is not a precise term. Very often, it means manufacturing industry only; nowadays it is frequently used to mean all kinds of economic activities, including services and mining. The usual way of *classifying* economic activities is:

1 Primary activities These involve natural products and resources and include farming, fishing, mining and quarrying (extractive industries).

2 Secondary activities These include all kinds of manufacturing industry. *Basic industries* convert raw materials into other products which are the raw materials for other industries (e.g. iron ore to steel). *Heavy industries* produce large, bulky products (e.g. structural steel, heavy chemicals). *Assembly industries* put together components made elsewhere (e.g. computers). There is an enormous range of types of manufacturing industry.

3 Tertiary activities These are services, and cover a wide range of activities. They include distribution of goods (i.e. wholesaling), retailing and transport; finance (i.e. banking and insurance); professional services (e.g. solicitors); public services (e.g. local and national government, education and health).

4 Quaternary activities These are concerned with research and development and the processing and handling of data. (Not everyone counts these as a separate group: they are often included in no. 3.)

Short questions or multiple-choice questions on this topic sometimes ask you to name examples of these, or to put named examples in the right category.

Industrial systems
These can be complex because:

1 A wide range of factors have to be considered when making decisions about locations, including: raw materials, markets, transport, labour, capital, power supplies, government and sites and buildings (see Section 11.5).
2 They operate on a world-wide scale. Raw materials and components come from many different sources, sales are in many different places and many specialist activities are involved.

Industrial regions
These develop because different industries and activities are linked together and benefit from being close together. In the past, these might have been industries using iron and steel and coal, which would benefit from a common location on coalfields. Modern industries which depend on up-to-the-minute research often locate in areas adjacent to universities and other research centres.

11

INDUSTRY
11.2 Factors of location

The *location of economic activities* is affected by the following factors:

1 Raw materials (whether e.g. bulky, low value or compact, high value materials).
2 Market (i.e. the area within which the product is sold).
3 Transport, particularly its cost and its effect on accessibility.
4 Energy (although in only a few cases is cheap energy a vital factor).
5 Labour supply, whether in terms of numbers or special skills.
6 Capital, or the availability of funds with which to set up the activity.
7 Government, either through direct aid or through regulations.
8 Technical developments, which change the effect of some of the other factors.
9 Historical background, where the present locations are the result of past decisions.
10 Site, or the availability of the actual land on which to locate.

You can think of these factors as having a *pull* on the location of an economic activity. Those which are more important for a particular activity have a greater pull or effect. The idea is best shown by referring to some examples.

(*a*) In the past, brewing was an industry found in every small town. Breweries were small because the *technology* of brewing meant that it had to be done on a relatively small scale at the *market* (i.e. the area where it was sold). Beer is a bulky product and it was only economic to *transport* it short distances in the past. More recently, breweries have become much larger. New technology has made it possible to produce very large quantities in a few large breweries more cheaply. Improved transport and, in particular, its reduced cost has meant that beer can be transported greater distances to market. The large breweries are located in or near the main centres of population, or the main markets, and transported to the smaller markets.

(*b*) Iron and steel-making grew up on coalfields where all the *raw materials* were available close together, thereby reducing very high transport costs. As transport improved and raw materials became exhausted, the industry tended to relocate at coastal locations so that raw materials could be imported by sea transport. General improvements in transport and technology eventually meant that a location close to markets would be an advantage. However, government decisions meant that new works were established in the old areas of steel-making on the coalfields, where there was a ready supply of *labour*.

(*c*) Several retail supermarket chains will only establish new store where there is a *site* large enough to provide parking space, since changes in

Factors of location

personal transport have meant that without parking for cars, many potential customers would not use a store. The site does have to be in or adjacent to a large town to provide the necessary market and labour force. Ease of access for supply lorries (bringing the raw materials) is also important, and transport is best suited to a location reasonably close to the motorway network.

The following three examples are based on developments in Britain:

QUESTION 11.1 Study the map below, based on a coalfield region of Britain.
(a) At which location would you have expected iron-making to have developed before the use of coal for smelting? Give reasons.
(b) At which location would you have expected the iron industry to have developed during the nineteenth century? Give reasons.
(c) At which location would you expect to find a modern iron and steel works? Give reasons.

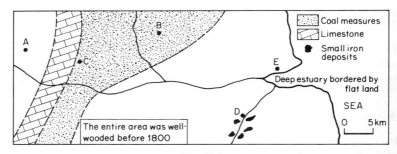

Fig. 11.1

QUESTION 11.2 What does the term 'market' mean?

QUESTION 11.3 In 1790, 8 tonnes of coal were need to make 1 tonne of iron. By 1850, only 4 tonnes were needed and in 1900 only 2 tonnes. Now, less than 1·3 tonnes is needed. Is this an example of a change in: (a) transport, (b) government influence, (c) power supplies, (d) technology?

11

INDUSTRY
11.3 Primary activities

Mining and quarrying (extractive industries) employ only a small proportion of the workforce. As the provider of basic raw materials and sources of power, their geography is important. Their location depends, first of all, on the distribution of the mineral to be extracted. That is fixed by geology. However, whether or not mineral deposits are exploited depends on several factors. You should know the following points:

1   Accessible deposits are exploited before less accessible, isolated ones.
2   As resources decrease, less accessible deposits are developed.
3   Larger deposits are exploited before smaller ones.
4   Richer-quality deposits are exploited before poorer-quality ones.
5   Scarcity encourages the search for new resources.
6   New techniques enable the exploitation of resources known but previously unworkable.
7   Easier mining conditions make development more likely.

These statements are generally true because of the low value of minerals in relation to their bulk, the importance of transport costs, and the very large amounts of capital (money) needed for new mining developments, especially in isolated areas.

Circumstances do change, and the geography of these activities can change as a result. For example:

1   The six-fold increase in oil prices in the 1970s meant that oil-drilling became possible even in very isolated and difficult areas, where costs had previously been too high. The North Sea and the North Slope of Alaska are two cases in point.
2   The development of large bulk carriers during the late 1950s and early 1960s meant that minerals could be transported much more cheaply and over greater distances. One result is seen in the iron-ore trade, with Australian ore being taken to Japan, the USA and even Europe.

Mining methods affect the location of mining. Depending on the type of deposit, mining can be more or less expensive. The main methods are:

Underground mining, where the mineral is found in lodes or veins, or, in the case of coal and other minerals which are really sedimentary rocks, in seams of varying thickness.

Open-cast mining, where the mineral is found very close to the surface and can be extracted by giant diggers once the overlying soil and rock (overburden) is removed.

Gravel dredging, where the mineral (generally a heavy metallic one) has been deposited over the centuries in river gravels.

129 *Primary activities*

Questions on this topic often present a theoretical example on which several problems are based. Alternatively, information from a particular mining area is given for you to analyse.

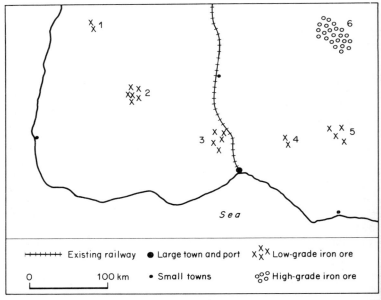

Fig. 11.2

QUESTION 11.4 Study the map above, showing an imaginary area of the world.
(a) Explain why the low-grade deposit at 3 was the first to be exploited.
(b) Of the other deposits, which is most likely to be exploited first? Give your reasons. Suggest what developments will be necessary.

QUESTION 11.5 Study the maps on page 130, showing coal-mines in north-east England at different times, and the geological cross-section.
(a) Describe the changes which have taken place in the geography of coal-mining in north-east England.
(b) Explain the changes you described in your answer to (a).
(c) What are the social and economic effects of the changes you have described?

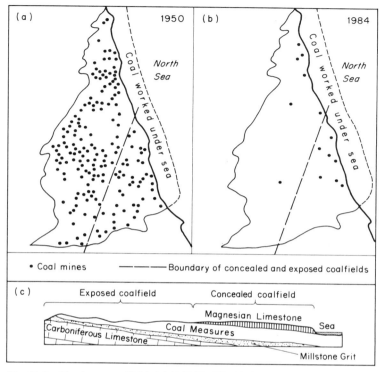

Fig. 11.3 *North-east coalfield: (a) 1950 (b) 1984 (c) Simplified geological section*

Consequences of mining and quarrying
The three important aspects are:

1 The effects on settlements and whole regions when mining declines or ceases altogether. When mining finishes, the only major sources of employment also disappears, and every other aspect of local economies is affected. In extreme cases, settlements decline as people migrate to other areas.

2 The impact of mining on the environment is drastic. It is most noticeable in areas where there has been no mining before, and in recent years there have been a number of planning inquiries into proposals to mine in new areas (e.g. the Vale of Belvoir). Questions are often linked to National Parks and the effects of mining or quarrying on those areas.

3 The exports of many less developed countries are heavily dependent on a small range of primary products, and in many cases these are minerals. It is common, too, for one mineral to dominate. For example, 75 per cent of Liberia's exports consist of iron ore, and in the case of Mauritania this is 93 per cent. Questions often use this as one characteristic of a less developed country and ask about the problems that result. These include the following:

(*a*) Exports are vulnerable to changes in world trade. A recession, or fall in trade, usually affects demand for raw material most of all. The result is a fall in exports and a drop in world prices, giving the exporting country a double blow.

(*b*) Competition from other producing countries also dependent on a limited range of exports keeps prices low and makes it difficult to forecast future exports.

(*c*) Prices of raw materials have consistently increased more slowly than prices of manufactured goods. Over a period of time, less developed countries need to export more and more to pay for the same quantity of imported manufactured goods. Oil exports have been the only real exception to this.

11

INDUSTRY
11.4 Secondary activities

Types of location
Industries which use large quantities of raw materials, especially bulky ones, generally favour a *raw material location*. Industries in which the costs of distribution are particularly important favour a *market location*. A *break-of-bulk location* is one where a change in transport occurs, usually at a port. A few industries use such large quantities of energy that they need an *energy location*. Many modern industries have no particular location requirement. These are called *footloose industries*.

Industrial systems
The kind of location favoured depends to a large extent on the industrial system. If you think of an individual factory as the centre of the system, the *inputs* will consist of raw materials or components, energy supplies, labour force, capital equipment and transport costs. The *output* consists of the production of the factory. This may be a finished product or it may be components which go on to another factory. Fig. 11.4 shows a simplified industrial system and represents in a basic way the workings of a company with one factory. Fig. 11.5 (page 133) shows a much more complex system: the company has several factories and processing plants and a comprehensive distribution system as well as a head office. As part of a bigger organisation, it is part of the European Region of Kodak which has its office in London. In turn, this group is responsible to the Eastman Kodak Company at Rochester, USA. For a single works within a big company, Fig. 11.4 would still be adequate, but with the addition of arrows showing decisions from the company's head office and any links with other factories in the company.

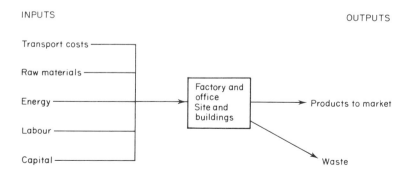

Fig. 11.4 *Simplified industrial system*

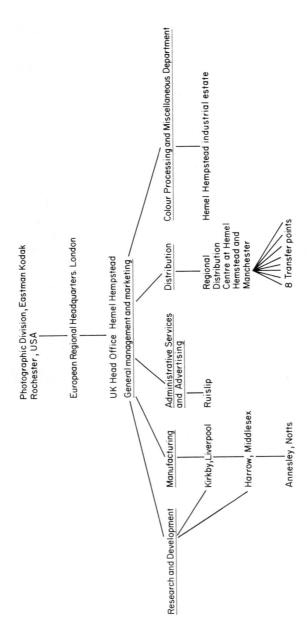

Only the vertical links are shown on this diagram. There are very many horizontal links between locations. If they were all shown, the diagram would be criss-crossed by dozens of lines. In addition, there are links with European establishments: factories research laboratories and distribution centres

Fig 11.5 *Simplified diagram showing the organisation and linkages of Kodak in the UK*

Decision making

The links in the chain of decision making are more complicated in a large organisation. As far as the geography of industry is concerned, the aspects of decision making that are most important concern the *location of industry*. The factors which affect decisions about locations are considered later in this section. However, there are some important points which are often overlooked.

1 *The original decision about where to locate* Often businesses are started in particular localities because that is where the individual entrepreneurs live. Such industries succeed if there are no special locational needs. A search for the best location is necessary where there are special needs (e.g. for industries using bulky raw materials).
2 *When a business grows* Different decisions have to be made, such as whether to: (*a*) expand the existing factory, (*b*) open a branch factory nearby or in another area, or (*c*) close the existing factory and establish a new, bigger one nearby or in another area.
3 *When trade is poor* Decisions have to be made, either to: (*a*) cut production, possibly closing a factory or factories, or (*b*) change to making different products.

Industrial links

The inputs and outputs of a factory make up a complex set of links with suppliers and distributors, possibly on a world-wide scale. In those areas where industries *cluster* because of favourable conditions, industrial regions develop.
2 With particular companies, different sorts of links occur. Large companies often increase in size by taking over others in the same business. This is called **horizontal integration**. Others expand by taking over their suppliers and distributors. This is called **vertical integration**.
3 The scale of operation is also significant. At one end, there are the small, locally-based factories which rarely operate outside their locality. They obtain their materials from local suppliers and sell to local markets. At the opposite extreme are the *multi-national* or *trans-national* companies, which operate right across the world. They are involved in most countries and although some, like General Motors, operate mainly in one industry, many are involved in an enormous range of totally unconnected activities. These companies are important because they are so large and wealthy and therefore have a greater effect on the economies of the countries in which they operate.

Industrial sites

All industries require land on which to build and questions frequently ask for comments on *sites*. Some kinds of industry can operate from converted barns or large houses (e.g. some small electronics firms). Their site requirements tend to be for a clean atmosphere, not specific buildings. Other industries, especially heavy ones, need large areas of land at relatively low cost and with firm foundations to bear the weight of heavy equipment. The site may need to be adjacent to a waterway, as in ship-building. Large concerns may also require a flat or easily-levelled site, a large works operating an assembly line needs everything on one level.

Questions on sites frequently use photographs, map extracts or large-scale sketch maps. You are likely to be asked to describe the characteristics of the site or to comment on its advantages and disadvantages.

Note that in cities in particular, industry is often found on sites which were undesirable for other uses and which may have been left clear at some earlier stage of growth. Thus, valley bottoms that were avoided by earlier settlement may now be occupied by industry.

Congestion and lack of space means that much industrial land at ports has been reclaimed from marshland or even from the sea. One of the best-known examples is the oil, petro-chemical and iron and steel industry on reclaimed land at the mouth of the Rhine at Rotterdam. Such reclamation for industry is only feasible if the *location* is right. In other words, the new site is only valuable if it is in a suitable general location.

Although site is one factor, there are some industries where the requirements are so limiting that very little choice is available.

Much large-scale industry has been located on reclaimed marsh and mud flats in crowded industrial countries and there have been major losses of wildlife habitats. New proposals have met with considerable opposition.

QUESTION 11.6 Study the sketch maps below, showing the sites of two works.
(a) What are the difficulties or disadvantages of site A?
(b) In what ways is site B a more favourable site?

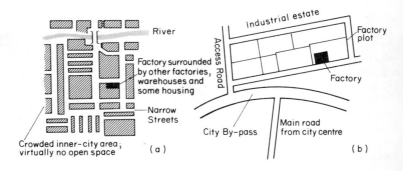

Fig. 11.6 *Factory sites*

Secondary activities 136

Industrial location

The types of location described earlier in this section follow on logically from the factors of location considered in Section 11.2. However, one factor is rarely all-important. You need to know some examples of particular locations to illustrate answers. As well as asking you to describe the importance of particular factors for a factory or industry you have studied, it is common to provide information about a location. Then you have to say why it is an attractive location for industry, and perhaps comment on any disadvantages.

Development areas: nowhere else comes within miles of Corby

If you're planning to develop your business you need look no further than Corby.

Corby is a **Development Area** so your business gets the help of Development Area benefits. For most companies this means the better deal for them of either 15% grants on plant, machinery and equipment or £3000 per job created. There is also selective assistance for some job creating projects.

Corby is also a **Steel Opportunity Area,** and this means even more incentives.

Corby is **England's first Enterprise Zone.** There are factories off the peg, from 500 sq.ft. to 50,000 sq.ft., some of which are rates free until 1991. You can also choose from offices, warehouses, and high tech buildings.

Corby has **EEC aid for small businesses.** £1m is now available to aid efficiency.

Above all, Corby is right in the heart of England. Within 80 miles of London. 50 miles from Birmingham. Strategically placed for any business that needs fast, inexpensive, easy access to the big South East and Midland population centres.

However far you look, you will find that, as a total package for the success of your business, nowhere else comes within miles of Corby.

Development Areas

as defined by The Department of Trade and Industry to take effect from 29.11.84

Fig. 11.7

137 Secondary activities

QUESTION 11.7
(a) Study the reproduction of a newspaper advertisement put out by the Corby Industrial Development Centre (Fig. 11.7).
  (i) How far is Corby from Birmingham and from London?
  (ii) Within what region of the country is Corby located?
  (iii) How much grant is available to companies for each job they create?
  (iv) What other financial incentives are there for companies to locate at Corby?
  (v) How does the advertisement itself help to promote Corby as a location for new industry?
  (vi) How does the advertisement suggest that Development areas are competing to attract new industry?
(b) For a *named* location you have studied, describe and explain its advantages as an industrial location.

QUESTION 11.8 Study the two maps below, showing the distribution of sugar beet factories and bakeries in the UK.
(a) Compare and contrast the distribution patterns shown by the two maps.
(b) For *either* sugar beet factories *or* bakeries, explain the pattern of distribution by reference to the relevant factors of location.

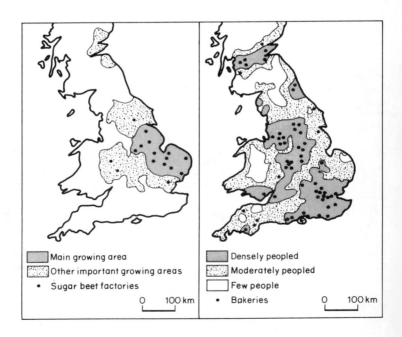

Fig. 11.8

11 INDUSTRY
11.5 Tertiary activities

These are services, and they are all *market* orientated. To a very large extent, locations are closely connected with the geography of settlements and central places theory (8.3). However, it is useful to have a summary of some points:

1 The ideas of threshold and range of a good mean that higher-order functions will be located in larger towns and cities, which, with their large spheres of influence as well as their own large populations, will have a big enough potential market.
2 Services are arranged in a hierarchy.
3 Accessibility is a key factor in the location of service activities.
4 Service activities include: the retail trade; the wholesale trade; office and financial functions; specialist services (i.e. those of use to particular industries as well as individuals). All are market based.
5 New developments in shopping and in office technology have brought about or are bringing about changes in some locations.

Distribution of services
Services are generally distributed in proportion to population, but there is an uneven element in the pattern because:

1 There is a concentration of office employment, including government, in London and the south-east in the UK.
2 Areas with greater levels of wealth, like the south-east, create a greater demand for services.
3 Services are concentrated in the larger centres within any one region.
4 Some other activities, like tourism, are themselves unevenly spread.

Office employment
As with factories, there have been changes in location of offices, in some cases. These changes have taken place in the UK as well as in other parts of the world, most notably in the USA. Due to the high cost of land in central London, together with high rates, congestion, time and cost of travel to work, a generally unpleasant environment and the fact that much routine office work could be done anywhere, many firms have either moved out of the capital altogether, or kept only a small head office there. In most cases, movement has been to other locations in the south-east. These new locations have been determined by: ease of communication with concerns in London; acceptability to staff; plenty of clerical and secretarial staff; availability of premises or land for new premises; site away from manufacturing industry.

This set of factors introduces another element into the question of location; that of *personal preference* or *personal perception*. In this case,

139 *Tertiary activities*

whether or not a location was acceptable to essential staff depended on their knowledge or image of a place – which might not necessarily be a correct image. This aspect of *decision making* is very important, because the people who make decisions about new locations are affected by them as much as anyone else.

QUESTION 11.9 Brent Cross is a modern shopping centre in north London, away from any established shopping centre. Study the map below and state five advantages Brent Cross has as a shopping centre.

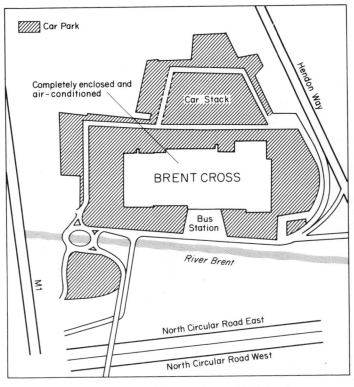

Fig. 11.9 *Brent Cross Shopping Centre, North London*

QUESTION 11.10 Study the map on page 140 showing the location of a major chain of department stores and the map on page 141 showing population distribution.
(a) Describe the pattern of distribution of the chain of stores.
(b) Explain the pattern of distribution of the chain of stores. Refer to the population distribution map and any other factors that are relevant.

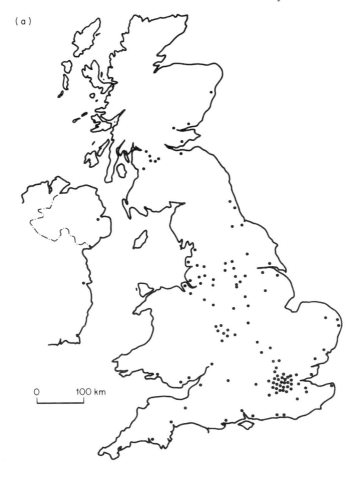

Fig. 11.10 *(a) Distribution of a chain of department stores*

141 *Tertiary activities*

(b)

Fig. 11.10 *(b) Distribution of population*

11

INDUSTRY
11.6 Industrial regions

Definitions
Concentrations of industry can be identified by plotting employment figures on maps and noting areas with above average numbers or percentages for industry in general, or particular industries. Alternatively, maps can be made to show the location of actual works.

Two types of industrial region are common:

1 Specialised regions, where usually old established industries persist, with a very limited variety of types of industry (iron and steel and textiles are common examples).
2 General regions, where a great variety of different types of industry are located, including many more modern ones.

On a map of a country showing both types of region, the former are likely to be on coalfields and the latter based on major cities and ports.

Growth and decline
All the developed countries of the world have areas where industry has declined. Conversely, there are areas of growth. The coalfield areas of Britain and the rest of Western Europe and the north-eastern states of the USA all provide examples of declining industrial regions. In Britain and Western Europe, governments have made efforts to halt the decline and to introduce new industries by a variety of kinds of *regional aid*, whereby firms receive financial and other inducements to locate in that region. (See the question on Corby in Section 11.4).

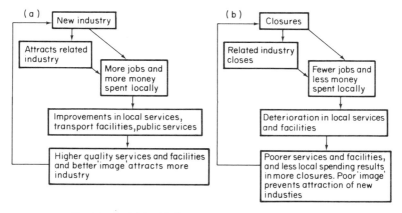

Fig. 11.11 (a) Multiplier model and (b) its reverse effect

Industrial regions

Questions on this topic ask for descriptions and reasons for the original growth and more recent decline of industry, in particular regions if they are specified in your syllabus or in a region of your choice. You should therefore study at least one example of an industrial region of this sort. Other questions are concerned with the mechanism of change. Thus Fig. 11.11 shows what is called the 'multiplier effect' or 'snowball effect'. The same model can be put into reverse to show the effects of closures.

QUESTION 11.11 Using the model on page 142, suggest the kinds of changes that would occur in the local area if the British Steel Corporation decided to close the steelworks at Port Talbot in South Wales.

QUESTION 11.12 For any long-established industrial region which has faced problems of declining industry:
(a) Describe the old industries and give reasons for their origins.
(b) Describe the modern industries and explain how they came to be located there.

Industrial growth on a regional scale is not very common in Britain, but growth regions occur for various reasons. During the 1970s, the North Sea oil industry brought about a period of rapid growth in and around Aberdeen, and in the south of the country, a 'corridor of growth' has been noted along the line of the M4 west from London. This is a zone where there has been marked growth in high technology industry as well as a variety of other light industries and service activities. Access to the motorway network and to London and its air and sea ports are all important factors. In addition, the pleasing environmental 'image' of the area is an attraction.

Core-periphery model

Many countries show a pattern of economic activity where one region is highly developed, relatively wealthy and having rapid growth. The rest of the country lags behind in some way or other. The developed region is the **core**, the rest of the country is the **periphery**. In the UK, the south-east would be the core and the rest of the country the periphery. In a less developed country such as Ghana, the region around the capital city in the south would be the core and the middle and north of the country would be the periphery.

The core-periphery model is more useful than just as a means of labelling or dividing a country into parts. In all sorts of countries, it is possible to see more than just two kinds of region; we can pick out the following:

1 The core, which has well-developed transport and communication systems, the bulk of the country's industry and service activity.
2 Declining zones (regions which are well established but whose industry is declining). In the UK, this could be a coalfield area or one of the rural uplands suffering rural depopulation.

Industrial regions 144

3 Growth zones, which are near the core and show rapid economic growth (e.g. the M4 corridor in the UK).
4 Resource-frontier zone, which is untouched country only now being occupied and developed. Many countries have such areas where there are small pockets of growth around resources.

The usefulness of models is that they help us to pick out general features which are common to many places and to separate them from the local features which make one particular part of the world different from other places.

QUESTION 11.13 Study Fig. 11.12, which shows a common pattern of development found in less developed countries.
(a) Suggest possible reasons for most economic activity being concentrated in one region of the country.
(b) For a *named* country or region, describe and explain the pattern of economic development. Refer to the model in your answer.

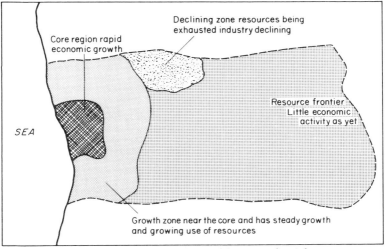

Fig. 11.12 *Economic growth model – core and periphery*

QUESTION 11.14 Study the maps on page 145 showing assisted areas in 1980 and 1984, and the maps showing job losses and levels of unemployment.
(a) Briefly describe the pattern of assisted areas in 1980 and 1984.
(b) Take each of the two following statements in turn and state whether or not they are true. (Give evidence from the maps to support your answer.)
   (i) The pattern of assisted areas reflects the pattern of the worst affected high unemployment areas.
   (ii) The south-east can be labelled the *core*, much of the rest of the country can be labelled *downward-transition zones* and a few areas can be labelled *upward-transition areas*.

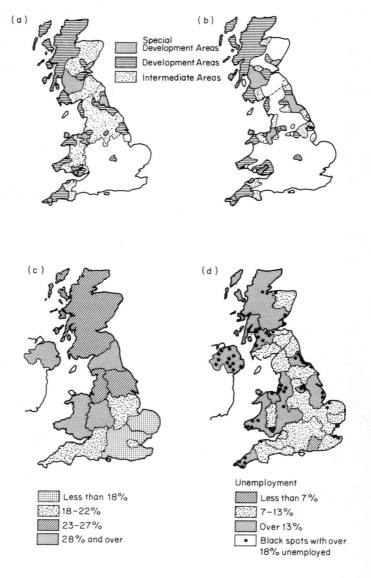

Fig. 11.12 *(a) Assisted areas 1980 (b) Assisted Areas 1984 (c) Job losses 1978–83 (d) Unemployment May 1984*

12

AGRICULTURE
12.1 Agricultural systems

Subsistence farming
This is mainly found in less developed countries. All or almost all production is consumed on the farm by the farmer and family. There are various types of subsistence farming:

1 *Intensive subsistence farming*, where small farm-holdings are cultivated to produce as much as possible. Any surplus is sold.
2 *Bush fallowing*, in which different plots of land are cultivated for a few years and then abandoned. They are not left for long enough for the natural vegetation, usually tropical forest, to regenerate. They are cleared and cultivated again while the area is still 'bush.'
2 *Shifting cultivation*, where virgin forest or regrown forest is cleared and cultivated for a few years before being abandoned once the soil's fertility is lost. This is found in areas where the population density is much lower than in bush-fallowing areas.
4 *Nomadic herding*, which is still found in North and West Africa, but is being squeezed out by a combination of environmental, political and population pressures.

Commercial farming
This involves the production for sale of animal or plant food products or industrial crops. There is a great variety of types of commercial farming:

1 *Plantation farming* is primarily located in the tropics and usually involves very large landholdings which concentrate on the production of one crop (e.g., rubber in Malaysia). The workers on plantations often have small plots of land on which they grow their own food. Many plantations are owned by large multi-national companies based in Europe and North America.
2 *Pastoral farming* mainly involves the rearing of cattle and sheep. It can be sub-divided into intensive and extensive types, differences being in the size of holdings and the effort involved.
3 *Arable farming* is the production of crops, of which there is a great variety. Intensive and extensive types can be distinguished.
4 *Mixed farming* is the most common and tends to be relatively intensive.

This division is useful on a world scale, together with a knowledge of the main areas of the principal food and industrial crops which can be found in any atlas, usually with the general climatic requirements. You should know in which climatic regions the major crops are grown. Smaller-scale studies (e.g. at the UK scale) need a more detailed classification of farm types.

Questions on farming systems very often make use of systems diagrams. The common elements are easy to identify: inputs, outputs and feedback.

147 Agricultural systems

Types of farming classification used by the Ministry of Agriculture, Food and Fisheries

1	Specialist dairy	6	Predominantly poultry
2	Mainly dairy	7	Pigs and poultry
3	Livestock rearing and fattening: mostly cattle	8	Cropping: mostly cereals
		9	General cropping
4	Livestock rearing and fattening: mostly sheep	10	Predominantly vegetables
		11	Predominantly fruit
5	Livestock rearing and fattening: cattle and sheep	12	General horticulture
		13	Mixed

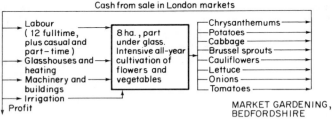

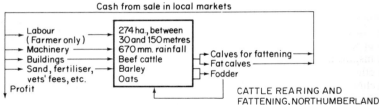

Fig. 12.1   *Examples of farm systems*

Questions on farming generally include the following types:

1   Short, often multiple-choice questions defining terms (e.g. arable farming, intensive or rotation).

2   Farm studies, either of a single farm or a comparison of two farms. Farm plans and other information are usually provided, and questions are concerned with the farm system and the pattern of land use on the farm.

3   Local area land-use patterns based on an area of anything from 5 to 50 km across. These questions are about the patterns of land use and the factors affecting the patterns. Explanations may be to do with physical factors, human factors or both.

4   Changes, problems and developments. This covers a wide range and overlaps other types. In particular, questions are about general agricultural and food-production problems, especially in less developed countries, and involve possible solutions and approaches to development. Depending on your syllabus, you would usually be asked to describe named examples.

12

AGRICULTURE
12.2 Factors affecting land use

How the land is used depends on the interaction of many factors. Physical factors set the limits of what can and cannot be done. Human factors (often referred to as 'human and economic') bring into being a different set of variations.

Physical factors
climate – temperature (maximum)
length of growing season
frost risk or months free of frost
rainfall total
seasonal distribution of rainfall
relief – height of land
slopes
drainage
soil – drainage properties
texture
fertility

Human factors
distance from markets or accessibility
transport facilities
methods of farming
technology
size of farms
availability of education and advice
banking and insurance
level of development
demand and competition
government influences

The importance of these factors varies, especially according to the scale of the study. On the large scale, climatic factors are especially important, but on a smaller scale others become important. For example, in the UK climate, relief and soils mean that the growing of crops on a large scale is only really possible in the east and south. However, factors like nearness to markets and size of farms and methods of farming have a great effect within the east and south on where market-gardening is located and where cereal-growing is located. Similarly, the fact that cereal-growing has replaced mixed farming on many large farms is due to government and EEC price support for cereals, making them exceptionally profitable crops.

QUESTION **12.1** For a farm or area you have studied, show how *physical* factors have affected the use of the land.

149 *Factors affecting land use*

QUESTION 12.2 Study the annotated maps (below) of two farms is different parts of England.
(a) Describe and explain how physical factors influence the land use on Cowbyers Farm.
(b) Describe and explain how human factors influence the use of the land on Manor Farm.
(c) Show two ways in which physical factors influence the use of the land on Manor Farm.

(a)

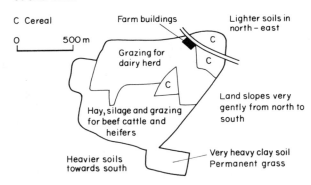

(b)

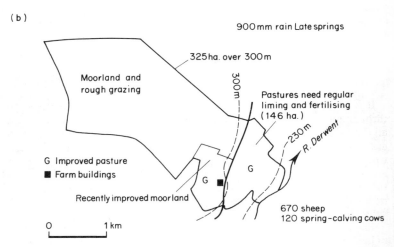

Fig. 12.2 *(a) Manor Farm, south Staffordshire (b) Cowbyers Farm, Northumberland–Durham border*

12

AGRICULTURE
12.3 Patterns of land use

Land-use models
The most common model of land use is based on the ideas of distance from market and intensity of land use. Fig. 12.3 shows a typical example of the way this model is presented.

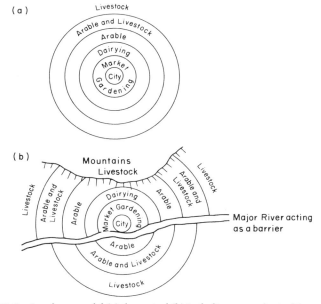

Fig. 12.3 *Land-use model (a) theoretical (b) including some physical features*

It is assumed that factors like climate, relief and soils are the same throughout the region and that transport is equally good in all directions. If all the farmers throughout the region produced the same mixture of crop and animal products, those nearest the market (the city on the diagram) would make more money because their transport costs would be lower. As a result, land near the city is worth more than land further away. To pay for the more expensive land, farmers near the city have to produce more. This means they have to farm the land more intensively and concentrate on more valuable products.

The rings shown on the diagram are labelled to give one possible pattern of land uses. Near the city are the more intensive land uses and further away land uses become less and less intensive. You can develop the model to include the effects of other factors.

151 Factors affecting land use

Questions based on the idea of this model are quite common. Note that the model itself may not be mentioned, although sometimes a diagram like Fig. 12.3 is included. The important point is that you understand how to (*a*) describe the patterns of land use, (*b*) identify relationships between patterns of land use and other features (e.g. soils), and (*c*) explain the patterns. Some questions are based on very general or even imaginary maps. Others use detailed information about specific places. Remember also that you may be asked to describe an example you have studied.

QUESTION 12.3 Study the map below, showing the distribution of different types of agriculture in Normandy, France.
(*a*) How does the map support the idea that land use becomes less intensive with increasing distance from major markets?
(*b*) What physical factors modify the effect of distance from market? Support your answers to both questions with examples from the area shown.

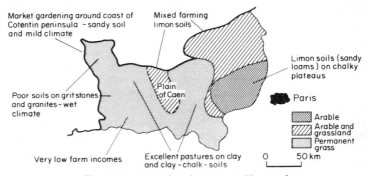

Fig. 12.4 *Agricultural patterns in Normandy*

QUESTION 12.4 Study the graph below, showing the profits from various farm enterprises at different distances from a major city.
(*a*) At what distance from the city would fruit and dairying be replaced by mixed farming?
(*b*) Draw a diagrammatic map to show the pattern of land use around the city.
(*c*) What would be the effect on the boundary between mixed farming and wheat production of a rise in the price of wheat?

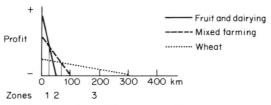

Fig. 12.5 *Profits from different types of farming around a city*

12

AGRICULTURE
12.4 Changes, problems and developments

Changes
Farming is a changing activity. In the UK, changes include:
1 A fall in the size of the work force.
2 An increase in mechanisation.
3 An increase in the average size of farm.
4 Increased yields.
5 A fall in the number of farms.
6 A greater degree of specialisation.
7 A decrease in the area of agricultural land as a result of competition from urban and industrial uses.
8 An increased use of pesticides, herbicides and inorganic fertilisers.
9 Changes due to government farming policies, particularly EEC (Common Agricultural Policy).
10 Increased involvement of companies from outside farming at the expense of individual farmers.

QUESTION 12.5 Describe and explain the changes in agriculture in an area you have studied.

Note that this question gives you plenty of choice. The scale of the answer is left to you, so that it would be possible for you to take anything from a single farm to an entire country. Whatever you choose, remember to name it and indicate its size. Also note that some syllabuses specify the areas you must study for particular topics, so make sure your examples come from the right areas.

Problems
Some problems are world-wide, some occur predominantly in less developed countries and some in developed countries.

1 Soil erosion is a world-wide problem (see 6.2). It is partly a direct consequence of farming methods, but in many areas it is a consequence of climatic and other environmental matters.
2 Water and its availability is a major problem in semi-arid areas and the development of irrigation schemes in less developed areas is hindered by lack of capital. The problem of salt accumulation has been mentioned earlier (Section 6.2). A further problem is that water for agricultural use is in competition with water for urban and industrial uses.
3 Water pollution, including ground-water pollution caused by chemical-rich water running off farmland or infiltrating into the soil and rock.
4 Reclamation of areas previously not farmed (e.g., moorland, floodplain meadows and marshes) removes the habitats of many species of plants and animals.

153 Changes, problems and developments in farming

5 Food-supply problems occur in many less developed countries (see below).

Questions on these issues tend to be part of question on topics such as environmental problems, planning or conservation, or problems of development.

Agriculture in less developed countries
Questions are about: (*a*) problems of agriculture, (*b*) solutions to the problems, and (*c*) patterns of land use and farming systems.

Problems
1 Environmental problems.
2 Population pressure.
3 Poor infrastructure (i.e. poorly-developed transport, banking, finance, advice services etc.).
4 Lack of technology.
5 Ill health and disease.
6 Landholding systems.

The importance of those problems varies from place to place, but population pressure is certainly a major problem in the majority of places.

Solutions
Solutions include specific programmes dealing with each of the problems. Since food production is the major difficulty for rapidly-growing populations, questions often concentrate on methods of increasing production, which include:

1 Increasing the area of cultivated land. This generally involves irrigation. It can be as part of a small village-based scheme, in which groups of people make small dams to store water from the wet season for the dry season, or it can be a large irrigation scheme like the Gezira scheme in Sudan. Large schemes, however, mainly involve the production of crops for export, and these are often industrial crops like cotton, not food crops. Other ways of increasing the amount of cultivated land include the reclamation and draining of marshland and the clearing of forest (see Section 6.4), which itself causes problems.
2 Increasing yields from land already cultivated. This can be done by a combination of approaches:

(*a*) Using new strains of plant.
(*b*) Pesticides and fertilisers.
(*c*) Reform of the landholding system.
(*d*) Improved storage methods.
(*e*) Improve threshing methods for cereals.

Changes, problems and developments in farming 154

(f) Irrigation.
(g) Improved transport and financial services.
(h) Development of agricultural advice services.

As with previous sections, you will need to study an example which can be used in questions that ask for your own accounts.

QUESTION 12.6 Read the description below of the changes on one farm in central England.
(a) In what way do the changes show that farming is becoming more specialised?
(b) What would be the effect of these changes on the number of farmworkers employed? Explain your answer.

Home Farm grew potatoes, sugar beet and cereals, and kept beef cattle, sheep and pigs. The whole operation was streamlined so that one farm with 417 hectares became an arable and beef cattle farm and the remaining 54 hectares were turned into a separate dairy farm. This dairy farm is run completely independently of the arable and beef farm. The reasons for this change were that the sheep had never been really profitable and to make a success of pig farming it is essential to specialise or give up.

Fig. 12.6 *Changes on a farm in central England*

QUESTION 12.7 Study the sketch map below, showing the land of one farmer in the Punjab village of Athaula.
(a) What *two* problems for the farmer are caused by the system of land-ownership?
(b) What other feature creates difficulties for the farmer?

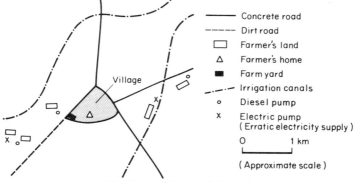

Fig. 12.7 *A farm in the Punjab*

155 *Changes, problems and developments in farming*

QUESTION 12.8 The 'Green Revolution' is the term given to greatly increased production largely due to the use of 'high-yield' varieties of rice, wheat and maize. Study the graphs below, showing changes in selected countries and the graph for inputs and outputs for cereal crops.
(a) What has been the effect of the use of new high-yield varieties?
(b) What does the second graph suggest is the vital element in the success of these high-yield varieties?
(c) Besides your answer to (b), what other inputs are essential to the success of the Green Revolution?
(d) In the light of your answers to (b) and (c), explain why developing countries, and particularly poorer farmers within those countries, are not benefiting to the fullest extent from the Green Revolution.

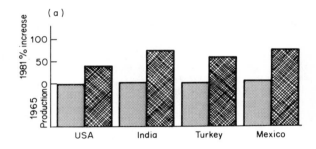

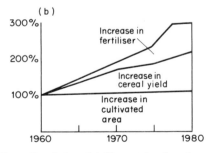

Fig. 12.8 *The Green Revolution: (a) Changes in selected countries (b) World-wide changes in inputs and outputs*

Changes, problems and developments in farming 156

QUESTION **12.9** Fig. 12.9 shows how a 90 000-person commune in China operates a virtually self-sufficient farming system and at the same time produces surplus food for sale in nearby cities.
(a) Explain how the farming system is intensive.
(b) Explain how levels of soil fertility are maintained without the commune having to buy mineral fertiliser from outside.
(c) How does the farming system enable the commune to be almost completely independent in energy supplies?
(d) How does the approach to farming contrast with that of the Green Revolution?

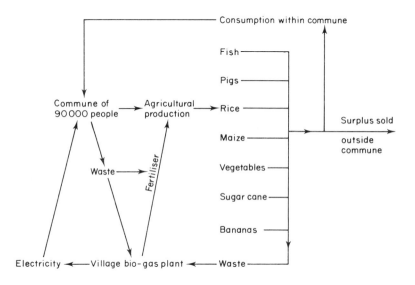

Fig. 12.9 *A commune in China*

RESOURCES
13.1 Types of resources

Resources are of many types and consist of anything that is of use. Usually, however, we mean *natural resources*. These include: minerals, rocks sources of fuel and power, timber, fish, soil, water, plants and animal life and even scenery.

Resources are grouped into two main sets:

1 **Renewable resources** or flow resources are those which are continuously available or which can be replaced or replenished. Sunlight or solar energy is continuously available. Water is also continuously available, although its availability can be disrupted by human activity or misuse. Timber, providing it is replanted or allowed to regenerate after felling, is replaceable in the long term. Fish too are replaceable.

2 **Non-renewable resources** cannot be replaced. Fossil fuels like coal and oil, and minerals like bauxite and iron ore can be used only once. After mining has taken place, reserves for the future are less. Of course, metals can be recycled.

The rate of use of the world's resources is a matter of concern in two ways:

1 Non-renewable resources have a fixed life at present rates of consumption and there have been predictions of total exhaustion of certain resources. However, different techniques or higher prices make it more economic to use poorer quality deposits or to extract more from the same deposits by reducing waste. Substitution of different resources is also possible.

2 Renewable resources which have been over-exploited can decline to the point where replenishment is no longer possible. This has happened with some species of whale. The herring fisheries of the North Sea declined to such an extent that a six-year ban had to be imposed in order to allow recovery. Another example of misuse of renewable resources is the causing of soil erosion by over-cultivation.

There is a major relationship between levels of development (Unit 14) and consumption of resources. The developed countries of the world have about 30 per cent of the population but account for 90 per cent of the consumption of raw materials, energy and manufactured goods. Since so many of the world's resources come from less developed countries, this shows that there is a flow of resources from the less developed to the more developed countries.

Types of resources 158

QUESTION 13.1 Study the sketch map below of a region in northern latitudes.
(a) Which resource is non-renewable?
(b) Name one resource which is renewable.
(c) Development of the area to exploit the non-renewable resource may well destroy other resources. Explain this statement.

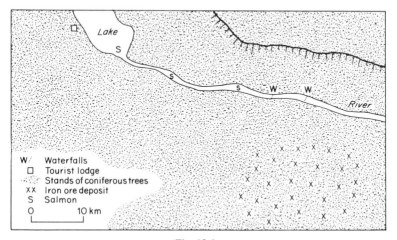

Fig. 13.1

QUESTION 13.2 Study the maps on page 159 showing the Kielder Reservoir in Northumberland.
(a) Kielder Water has been created on the headwaters of which river?
(b) What resource was lost in the creation of this lake?
(c) The reservoir was created to provide water for the main urban and industrial areas of the region. How is the water transferred?
(d) The Kielder Dam and Reservoir Scheme serves another purpose besides water supply. What is its other purpose? Give examples of the different ways by which it achieves this other purpose.
(e) What is a multi-purpose scheme? Describe an example of a water-based scheme you have studied, and explain how it achieves its aims.

159 *Types of resources*

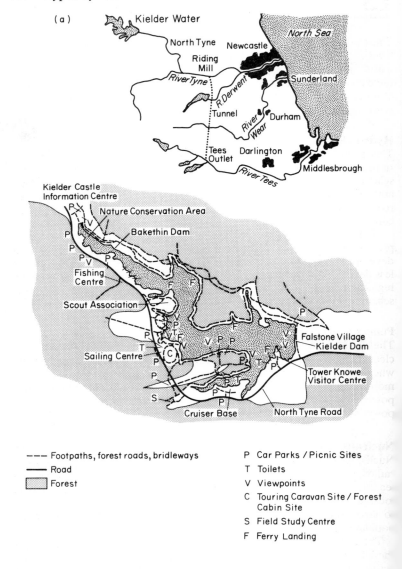

Fig. 13.2 *Kielder Water, Northumberland*

13

RESOURCES
13.2 Energy resources

These are particularly important, and there is a direct relationship between level of development and energy use. Thus the USA with six per cent of the world's population consumes 28 per cent of the world's energy. Aspects of the exploitation of fossil fuels have been covered in Section 11.3, but you also need to consider the development of *hydro-electric power* and *nuclear power*.

Hydro-electric power (HEP)

This is renewable and developed widely. The advantages are that it has low running costs, is pollution free and its costs can be shared with other schemes like irrigation, recreation, water supply and flood control. Disadvantages are high initial capital costs, flooding of farmland, trapping of river silt and the effect on the landscape. (Note that the last might be considered a gain by some.)

HEP is frequently located in isolated areas, since mountain regions provide both high water supply and the fall necessary to generate power. In developed countries, most HEP potential has been put to full use, but in less developed countries its use has been more limited, due mainly to the high capital costs. Notable exceptions are south-east Brazil and individual schemes like the Aswan High Dam in Egypt.

Pump storage schemes

These use surplus power from conventional power stations (coal or nuclear) to pump water into high-level storage lakes. At times during the day when demand for electricity rises, this water is released to provide HEP more or less instantly. Some such schemes are based on National Grid power, others have been based on paired power stations (e.g. a nuclear power station and HEP station immediately adjacent to each other).

Nuclear power

Nuclear power stations need only small amounts of fuel, but the disposal of radioactive waste is a major problem. The risk of accidents and radioactive leaks is also a significant factor. Earlier nuclear power stations tended to be located in relatively isolated places, with solid rock foundations and access to large amounts of cooling water. Many are now located in densely populated areas, especially in the USA and Europe.

The explosion at the Chernobyl nuclear power station north of Kiev in the USSR in 1986 affected places as far away as the British Isles. Radioactive contamination in the Chernobyl area resulted in the permanent evacuation of villages and farmland. This disaster and other near-accidents may affect future policies on the development of nuclear power.

Energy resources

Coal-fired electricity power stations

These have two sorts of location. Older ones tend to be located in major population centres and rely on coal brought in by water. Nowadays the majority of these power stations are located on coalfields, and the electricity generated is fed into the National Grid for transport by power line to other areas. Favoured locations within the coalfields are along rivers, which provide cooling water.

Alternative energy sources

These are less well developed and some are still in their very early stages.

Tidal power Here, the rise and fall of the tide is used to drive turbines and produce electricity. Even compared with HEP, construction costs are extremely high. Possible sites are limited by the small number of places with the necessary large tidal range.

Geothermal power This uses underground water heated by volcanic activity to produce steam to drive turbines. Electricity is produced in this way in New Zealand and Iceland. Again, possible locations are limited.

Solar power On a large scale, this requires expensive equipment at present, as well as a reliable climate. On a local and domestic scale, it has been proved useful as the sole source of heating or as a supplementary source of water heating.

Wind power At present, this is also more significant at the local scale. It is used to pump water and provide electricity in isolated places.

Conservation

Conservation of energy involves making more careful use of energy resources so that they last longer and so that less energy is wasted. There are many methods of conserving energy:

1 Increase in the efficiency of power stations to convert more of the energy from the coal or oil into electricity.
2 Use of waste heat in housing or greenhouse schemes.
3 Use of domestic and industrial waste as fuel.
4 More efficient industrial processes that will produce more using less energy.
5 Insulation in buildings of all types to cut heat loss.
6 Better initial design of buildings to conserve heat.

Energy resources 162

QUESTION **13.3** Study the pie charts below, showing changes in the sources of energy used in the UK.
(a) Which energy source showed the greatest relative decrease in importance?
(b) Explain why there has been a great increase in the importance of natural gas.
(c) Name and explain one other change shown by the graphs.
(d) In recent years, there has been a fall in *total* energy consumption. Suggest reasons for this change.

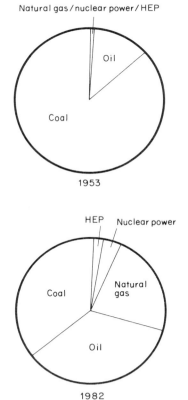

Fig. 13.3 *Sources of energy used in the UK, 1953 and 1982*

RESOURCES
13.3 Conservation of landscapes and wildlife

13

Questions on this topic tend to concentrate on National Parks and other such areas which have been labelled as areas of special scenic, wildlife or scientific value, and planning conflicts over proposed new developments.

The need for protection or conservation has arisen because of increasing population pressures and, in particular, greater use of areas for leisure purposes. This is due partly to increased incomes and leisure time and partly to greater ease of access by road (greater car-ownership and increased accessibility from motorways). Other land-use pressures in the form of farming, forestry, quarrying and housing also change landscapes and must be taken into account.

The UK is fairly typical of developed countries, especially the smaller, more crowded ones. The following list covers most of the developments related to protecting the environment, and more particularly the rural environment, as well as catering for the increase in outdoor leisure activities.

National Parks
There are nine National Parks in England and Wales. Their purpose is to conserve and improve access to those areas of particularly great scenic beauty. Note, however, that the land is not owned by the nation: the bulk is privately owned and people in National Parks have to make a living, whether from farming, forestry, quarrying, tourism or any other activity. There are tight planning controls, however, and new developments are restricted. Inevitably, conflicts of interest arise (e.g. quarrying in the Peak District, the route of the Okehampton by-pass on the edge of Dartmoor, provision of water supply in the Lake District). Tourists themselves cause problems, especially in places where large numbers congregate.

Areas of Outstanding Natural Beauty (AONBs)
These have a different distribution from National Parks, which with the exception of the Pembrokeshire Coast National Park are all in high or mountainous regions. The AONBs are mainly hill areas in the south of the country. They receive benefits in the form of additional grants and protection, but not to the same degree as the National Parks.

The National Trust
This organisation owns fairly large amounts of land, mostly in small patches throughout the country, including much coastline.

Country Parks
These have been created in most districts of the country. They offer open air, scenic views and nature trails, providing leisure facilities nearer the

Conservation of landscapes and wildlife 164

main population centres. Their individual attraction varies according to the site that was available.

Nature Reserves and SSSIs
National Nature Reserves and Sites of Special Scientific Interest are distributed around the country. They generally occupy small areas of land.

Questions on this topic follow a typical pattern of either providing information about a particular area with questions to be answered, or asking you to use an example you have studied. The question may be straightforward, describing the attractions or importance of an area, or it may be centred on a planning issue.

Other countries have different polices with regard to conservation. In the USA, the National Parks are very large and are owned by the nation. Unlike those in England and Wales, they have only two functions: conservation and recreation. Even so, there are still problems in a few of the most popular spots (e.g. Yellowstone National Park).

Less developed countries have also set up National Parks. Again, the range of functions is limited, usually to conservation and tourism (e.g. in Kenya, large numbers of tourists visit the National Parks every year to see the wildlife). However, there are pressures from two sources: population pressure, as potential farmland is used up, and pressure from poaching by hunters. The limited resources of the National Park services are therefore stretched in trying to protect animals.

QUESTION 13.4
(a) There are ten National Parks in England and Wales. (*i*) What is a National Park? (*ii*) Name two National Parks and say how their landscapes differ.
(b) Fig. 13.4 is based on an area in one National Park. There is a proposal to build a reservoir. (*i*) What would be the effects of building this reservoir? (*ii*) Name two groups of people who are likely to *object* to the proposal, and two groups who are likely to be *in favour* of it. For each group in turn, explain the attitude to the proposal.

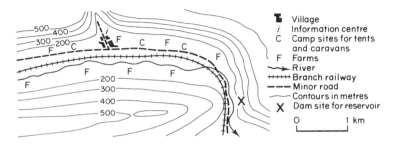

Fig. 13.4

CONTRASTS IN DEVELOPMENT
14.1 World patterns

14

The term **development** covers all aspects of the economic life of a country. A range of *development indicators* can be used. Sometimes *wealth* is used. For this purpose, gross national product (GNP) per head is calculated by dividing the value of all goods and services produced in a year by the total population. On this basis, a rough comparison can be made. Thus the GNP per head of the USA is 40 times that of India. By itself, wealth gives only a partial guide to development. For example, GNP statistics for some oil-exporting countries are extremely high, but this is not necessarily reflected in the standards of living of the people, since there may not have been time for this wealth to be translated into an improved quality of life. In addition, one set of figures hides the fact that averages do not show the way wealth is distributed within a country (e.g. 90 per cent of the wealth could be in the hands of 10 per cent of the people).

Indicators of development

Indicator of development	Developed Country (DC) UK	Less developed Country (LDC) Indonesia	Least Developed Country (LLDC) Ethiopia
Population growth %	0·1	2·3	2·0
Birth rates ‰	13	35	48
Death rates ‰	12	13	24
Infant mortality rates ‰	12	105	145
Life expectancy (years)	74	54	46
People per doctor	650	11 530	58 490
Calories per person per day	3306	2315	1735
Urbanisation (%)	91	21	14
GNP per person ($)	9110	530	140
% working in agriculture	2	55	80
Literacy rate %	99	62	15
% attendance at secondary school	82	28	11
Energy consumption (000s tonnes of coal equivalent)	311 327	118 218	71
Telephones per 100 people	49·7	0·4	2·8

The table above lists possible indicators of development. You would not expect to be given such a long list in an examination question, but you should be aware of their meaning and know what typical figures are for different kinds of countries. This enormous range of indicators is far from complete, but includes the ones most likely to be used in an examination question. They could be presented in a table, as here, or as a graph, in a piece of text or in diagram form. They could even be hidden in a cartoon – not as figures but as ideas and meanings. No matter how material is presented, be prepared to look carefully. There is a strong *correlation* between the indicators in the list and any pair of them could be used in a *scattergraph*.

Developed countries (DCs), and less developed countries, (LDCs) have been mentioned frequently in this book. Sometimes, the poorest countries are referred to separately as least developed countries (LLDCs). On almost all the indicators, they come off worst.

Many different sets of terms have been used for the poorer countries of the world. They have included: underdeveloped countries, developing countries, less developed countries, the Third World, and most recently, the South (the developed countries being called the North).

Many of the features of less developed countries have been dealt with under individual topics, particularly population, urbanisation and agriculture. Depending on your syllabus, you may be able to use examples from many different places or you may have a more restricted choice. On the world scale, the distribution of countries makes a simple pattern. North America, Europe, the USSR, Japan, Australia and New Zealand make up the developed world and the bulk of Latin America, Asia and Africa make up the less developed world. Most of the least developed countries are in Africa, in the regions with severe environmental problems.

There are distinctions amongst the less developed countries, between what are called middle-income and low-income countries. Some have developed considerably in the last 20 years. A small number of *rapidly industrialising countries* have changed enormously, but not without problems. These include Hong Kong, Singapore, Taiwan and South Korea. Other large countries like Brazil have developed industry based on large foreign loans. Repayment of these loans, coupled with extremely high interest repayments, have created enormous problems.

Problems of development

There are no simple, single solutions to the problems of development. Some of the problems and their causes have already been included in the unit on agriculture, but they and others will be listed more fully here. Not all the features in this list apply to every country.

1 Population growth High population counteracts advances made in other areas (e.g. continued population growth creates increasing problems in providing new schools and training teachers). At the other end of the age range, greater numbers of old people create demands for other services. In between, there is the need to create more and more jobs and to provide more and more houses. Economic growth is therefore necessary to keep pace with population growth. Reductions in the birth rate are important. Although there have been dramatic falls in birth rates in much of Asia, progress in Africa and Latin America has not been so encouraging.

2 Ill-health and disease Problems resulting from undernourishment and malnutrition produce deficiency diseases and people are less able to withstand infectious diseases. Lack of money also means that people are unable to buy the medicines which would prevent other diseases being contracted.

3 Low literacy levels and poorly-developed education systems At the most basic level, these create barriers to the spread of information and new ideas. At other levels, they create a shortage of the skills needed to develop new industries or to introduce new methods.

4 Unbalanced population structure This is closely linked to population growth. Very rapid growth rates mean that a large proportion of the population is young (e.g. in Pakistan 45 per cent of the population is below the age of 15, compared with only 21 per cent in the UK). Some of the associated problems have been mentioned above under 'population growth'.

5 World trade structure Less developed countries depend mainly on primary products for their exports (i.e. agricultural and mineral raw materials). Their imports tend to be mainly manufactured goods. In the long run, prices of primary products rise more slowly than prices of manufactured goods. The result has been that less developed countries have needed to produce and export more and more primary products to pay for the same amount of manufactured goods.

6 Environmental problems Many less developed countries have specific environmental problems which hinder development. Semi-arid areas suffer from unreliable rainfall, making agricultural development difficult unless water supplies can be assured. Tropical rain forests have heavy rains which rapidly erode the soils if the forests are cleared. In addition, the soils of such areas are generally poor, so they have little potential for agriculture. Periodic hazards like hurricanes can devastate areas. Tropical areas are also prone to many diseases which affect humans and animals.

7 Lack of technology The lack of *appropriate technology* limits the progress that can be made (e.g. new seeds may give greatly increased production, but without the means to store the food properly, large amounts can be lost). Hand- or animal-operated machinery for threshing results in smaller losses than traditional methods but it still needs a labour force; whereas high-technology equipment would lead to fewer jobs and create maintenance problems because of a shortage of the necessary skills for maintenance.

8 Poorly-developed transport and communication The lack of all-weather roads or limited development of railway systems often limits other developments. The difficulties involved in exporting and importing materials are sometimes immense, particularly for landlocked countries. Transport systems are usually better developed around capital cities and this tends to concentrate more and more development in such regions, so that more isolated regions fall even further behind.

9 Lack of capital and high interest repayments Major developments, particularly of the **infrastructure** (e.g. transport systems), need vast amounts of money or capital. The problems of raising capital from exports is clear (see point no. 5 about world trade). The other source of capital is to borrow. However, even where generous lending terms have been agreed, problems have resulted. Where normal lending terms have applied, the results have been disastrous for many countries. High interest rates have meant that further borrowings have been needed to pay interest on the first loans. Under these circumstances, development becomes almost impossible.

10 Unfortunate effects of some aid Aid is often tied to specific purchases from the 'donor' countries, and such purchases may not necessarily be in the long-term interests of the country concerned. Actual gifts of commodities may undermine developing activities (e.g. large consignments of free powdered milk totally destroyed a project to develop a dairy industry in one area of a West African country). In the past, aid was often tied to a major project for which a high level of technology was needed. Such projects concentrated development in particular areas.

11 Lack of natural resources, especially energy resources Resources are unevenly distributed, and although some countries benefit enormously from their good fortune in having valuable resources, the majority are not in such a situation. The cost of developing many resources is so great that less developed countries have to depend on multi-national companies in order to develop their resources. The result is that much of the wealth produced leaves the country. Large amounts of capital are needed to develop energy resources, especially hydro-electric power, and the problems of lack of capital have already been examined in point no. 9.

12 The gulf between rural and urban dwellers Political power tends to be concentrated in the cities and decision-making is influenced by the needs and demands of the dense concentrations of people living in the cities. One result is that policies often favour the cities, to the disadvantage of rural dwellers. For example, the prices of food are sometimes kept low to satisfy town-dwellers, leaving the farmers with no incentive to produce more or to try new developments.

Approaches to development

Over the years, different solutions for dealing with development problems have been put forward. Amongst these have been:

1 Development of major projects to act as a stimulus to further development These depend heavily on foreign aid and in some cases they were

World patterns

clearly seen as status symbols (for the aid-giving country as well as the receiving country). This sort of project needs a high level of technology as well as capital (e.g. the Volta Dam Project in Ghana).

2 Development of industry to absorb and provide work for the rapidly growing population Such industry tended to be located in the *core region*, in and around the capital city. It often acts as an extra attraction to migrants from the countryside.

3 Processing of exported raw materials The value of processed or manufactured products is many times greater than that of the raw material from which they are made. Some countries tried to put pressure on foreign companies to locate at least some of their processing there, in return for being allowed to develop the resources (e.g. the processing of bauxite into alumina in Jamaica).

4 Large-scale rural development projects These are often linked to irrigation schemes, and mainly produce large amounts of export crops. Being susceptible to changes in world prices, these projects have not all been successful, and have had the added disadvantage of reducing actual food production.

5 Small-scale rural development Limited help and advice have been given on a village scale: the inhabitants then undertake all the work and make all decisions themselves.

6 Infrastructure development By concentrating on developing roads, banking and other services, the conditions necessary for development are provided.

7 Development of low-technology, labour-intensive industry rather than high-technology, capital-intensive industry and the use of imported manufactured goods.

8 Tackling environmental problems by the adoption of soil conservation approaches.

9 Reducing the imbalance between rural and urban areas (whereby the latter are favoured and the former forced to accept low prices for their agricultural produce).

10 Increasing international aid generally

11 Reducing international aid

This gives an idea of the opposing views that exist on how the problems of the poorer countries, and particularly the poorest countries, of the world should be treated. Note, however, that circumstances and problems come in such a great variety that:

(*a*) There is no best solution that would work everywhere.
(*b*) One solution by itself would not be enough, since all aspects of the geography of the countries are interrelated.

Questions on this topic often concentrate on particular projects, however large or small, and consider their aims, advantages and problems. You should be familiar with examples at different scales and from the areas specified by your own syllabus.

QUESTION 14.1
(a) Name three features which might be used as indicators of the 'quality of life' in a country.
(b) Suggest reasons why the urban populations of less developed countries are rising rapidly.
(c) With reference to a specific country, explain why problems often arise where a country depends heavily on one raw material export.

QUESTION 14.2
(a) Why do a number of less developed countries seek to promote tourism?
(b) With reference to a country you have studied, consider the advantages and disadvantages of the growth of tourism.

CONTRASTS IN DEVELOPMENT
14.2 Contrasts within countries

14

The same measures that were used in the last section for distinguishing between countries can be used within countries, within regions and even within cities. The broad division of countries into developed, less developed and least developed hides marked differences. At the scale of the EEC, for example, there are differences between countries, especially between those of northern and southern Europe. At the regional scale in the EEC, certain regions stand out as above or below average. The Mezzogiorno (southern Italy) is noted as a poorly-developed region with high unemployment, high percentages in agriculture, low value of production and low incomes. At this sort of scale, the Mezzogiorno shows up as part of the periphery of the EEC when it is viewed using the core-periphery model.

QUESTION **14.3** Study the graph below, showing GDP per person for the EEC. Which countries seem to belong to the core and which to the periphery of the EEC?

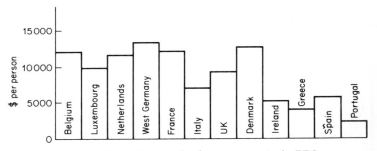

Fig. 14.1 *Gross Domestic Product per person in the EEC*

Within less developed countries, differences are even more marked than within developed countries. This is because the differences are accentuated by the **dual economy**. This means that one part of the country (the core) has an economy based on modern industry and modern methods, needing large amounts of capital, and for the rest of the country the economy is based on a combination of subsistence agriculture and craft industry. The idea of development planning was that the growth in the core region would gradually spread to other parts of the country. This was called the 'trickle down theory'. In reality, there has not been enough wealth to create the demand for industry's products outside the core.

Contrasts within countries 172

QUESTION 14.4 Study the map below, based on a less developed country.
(a) On your own copy of the map, shade and label the core region of the country.
(b) Outside the core, there is one area in which rapid development is taking place.
 (i) Mark and label the area on your own copy of the map.
 (ii) Explain why development is taking place in the area you labelled in (i).

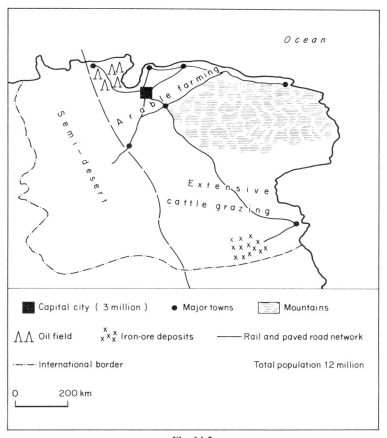

Fig. 14.2

REGIONS 15

(*a*) A region is an area of the earth's surface which has certain common features throughout. Many regions are identified on the basis of one set of features (e.g. an industrial region, an agricultural region, a climatic region or a relief region). Other sorts of regions are focussed on a particular place, like a port's hinterland or a city's sphere of influence. Sometimes the boundaries of countries, together with the official boundaries within countries (e.g. county boundaries) are used to define regions.

The term **geographical region** can include any of the above. In its most complete sense, it covers all aspects of the geography of an area, showing how each different element interacts to make the region distinctive.

(*b*) Your syllabus may specify particular regions to be studied. In most cases, this is done on a very broad scale, and the proportions of work to be based on the British Isles, other developed countries, less developed countries, and the world as a whole unit are specified. There is a very wide choice of examples. In some syllabuses, however, you are given very precise regions from which you must take the examples for particular topics. *Check your syllabus*.

(*c*) Although syllabuses are largely set out in topics, some questions are bound to involve ideas from several topics. For example, questions on the distribution of population in any country will also involve ideas about climate, relief, soils, agriculture, industry and settlement. In such a question about Scandinavia, for instance, you would have to show the relationships between relief and geology, climate, soils, natural resources, agriculture, industry and communications and the distribution of population itself.

Questions which involve this approach are likely to be of two types:

1 Map-reading questions often ask for a description of an area shown on the map extract or a comparison of two areas. In such a small-scale study, you still follow the same systematic approach. If no guidance is given in the question, follow a logical order and work through relief and landforms, geology (if appropriate), surface drainage, any apparently natural vegetation, land use, settlement, transport and communications. When asked to compare or contrast areas, do *not* write two separate descriptions. Instead, write about both areas under each topic heading, pointing out the differences and/or similarities.

2 On the larger scale, you could be asked to compare two areas of a country or to describe one and explain the relationships between certain features.

QUESTION 15.1 Study the Ordnance Survey map extract. Compare and contrast relief and land use in squares 2205 and 2506.

Regions 174

Since few syllabuses specify regions for study, the question is likely to provide information for you to interpret. If it does not, you should answer a question on a particular topic using a region you have studied as the example.

QUESTION 15.2 Describe the features of glacial erosion to be found in a region you have studied.

QUESTION 15.3 Study the maps provided (below and overleaf) showing the distribution of population and natural regions in the USSR.
(a) Describe the distribution of population within the country.
(b) Comment on the relationship between population distribution and the pattern of natural regions.

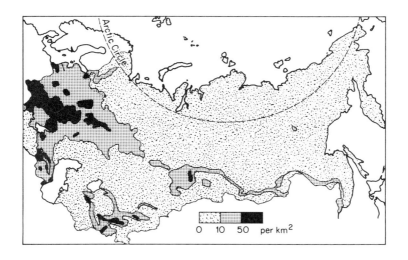

Fig. 15.1 *The USSR: population*

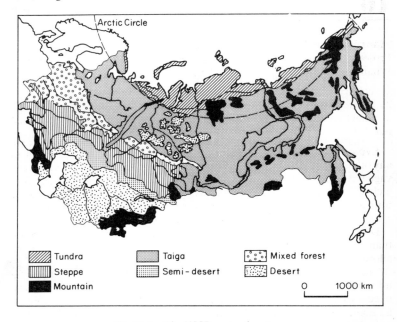

Fig 15.2 *The USSR: natural regions*

16

MAKING THE GRADE
16.1 Projects

Planning a project

Projects of various sorts form part of the course work of most syllabuses. Whatever words are used, your tasks are to:

1 collect information needed to answer a question or solve a problem;
2 present your information in a suitable way;
3 analyse and interpret the information to provide an answer to the initial question.

Every project involves fieldwork in a small area near your home or in any other convenient location. Every project should be specific. Do not use vague titles like 'canals', 'glaciation' or 'traffic problems'. Turn the title into a clear question like 'How is the Shropshire Union Canal being used for tourism?', 'What features of glaciation can be observed in the Fort William area?' or 'Why are there traffic problems in the city centre?'

Do not be too ambitious. You will not normally have to write much more than 2000 words, so the project must have clear limits. (You may be allowed to do two or three smaller projects instead of one larger one.)

The other practical point is to make sure you are able to collect the information. Can you get to and around the area of your study easily? If you need to keep to a set timetable, avoid clashing with other events like a holiday. Also remember that other subjects will have course work and you must give them all a share of your time.

Steps to follow

(a) Preparation

1 Decide on your project title or question.
2 Break down your main question into smaller questions.
3 Work out what information you need to collect.
4 Decide when and how you are going to do the fieldwork and *stick to your plans*.
5 Decide how to record your information.
6 Decide how you are going to present your information – in maps, diagrams, tables and photographs as well as writing.

(b) Fieldwork and follow-up

1 Set aside the time and get the fieldwork done without a break (except where your question demands that information should be collected over a long period of time).
2 Write rough versions of all your work. You cannot arrive at a final version without doing at least one draft version and possibly more.
3 Set out your work clearly, neatly and fully; make it easy for your external assessor to follow all your steps.

Projects

4   The most important part of your work is the analysis of information and your conclusions. This is usually worth half the marks for the project, so spend plenty of time on it.
5   The time you spend should reflect the importance of the project. If it is worth 30 per cent of the marks, your teacher should make allowance for time to be spent on the study. Remember, though, that normally the fieldwork will have to be done entirely in your time.

It is important to remember that, whatever your project, you should *discuss it fully with your teacher*. You are not expected to do the entire project without any guidance.

Examples
Every study is different, and here are examples which show how two students followed the steps above to produce excellent studies.

How does the hydrological cycle differ between two drainage basins?
The study compared the movement of water through the hydrological cycle in two local streams, one with a rural catchment and the other an urban catchment. To do this, it was necessary to have a thorough grasp of the working of the drainage basin hydrological cycle.

Data collection involved: measuring rainfall in the area; measuring stream flow in both catchments; making detailed land cover maps of both catchments; studying soil and geology maps.

Follow-up work involved: comparing the two catchments according to their actual stream flow and the speed with which water moved through the system; describing and explaining the differences between the two catchments and the relationships between rainfall and stream flow.

How will the new International Centre affect the town of Bournemouth?
This study described the building of the centre and asked the questions:

Will the centre help increase the attraction of Bournemouth?
How many people visit the centre and why?
Will it take business away from the Pavilion, Town Hall and the Pier Theatre?
Will it attract business from the other conference centres?
How does the public view the Centre?

These questions were answered by: conducting various surveys of visitors to the Centre and residents of the town, and mapping functions in the centre and elsewhere.

To give an idea of the variety of possible projects, here are just some of the many you could do:

1   How and why does the soil vary along a slope?

2   What is the effect on local weather of a frontal depression?
3   How is rainfall and stream flow related in a particular catchment?
4   Is there a pattern to the distribution of crime in a particular town?
5   How do the market areas of two local shopping centres differ and why?
6   Is there a pattern to the distribution of house prices in the town area, and if so, how can the pattern be explained?
7   How have the villages in an area changed in the last 20 years?
8   How does the quality of the environment differ between two areas of a town?
9   How has the pattern of industry in a town changed and why?
10   Are there distinct functional zones within the CBD (Central Business District)?
11   How do different groups of people perceive their neighbourhood?
12   How could the environment of the town centre be improved?

MAKING THE GRADE
16.2 Preparing for the examination

16

Well before the examination you should check carefully how many papers there are, what sorts of questions are asked in them, how many you have to answer and how much time you will have. If you are allowed to use an atlas in the examination, make sure you become thoroughly familiar with it. Make use of it frequently throughout the course. Look at examples of actual or specimen examination papers and practise answering the different types of questions in the set time.

Types of questions
Examples of all sorts of questions have been used throughout this book. Here are some more examples of each type. The short questions usually come on a separate paper from the longer ones.

Short questions
1 Questions demanding brief or even one-word answers, dealing with definitions of terms and simple facts:

QUESTION **16.1** What does the term 'nodality' mean?

QUESTION **16.2** Into which sea does the River Tyne flow?

2 Multiple-choice questions which can deal with your knowledge of facts and terms and with explanations:

QUESTION **16.3** Which *one* of the following is a feature of central business districts: high land values, open space, industrial estates, high-rise flats?

3 Alternatively, information may be presented in the form of tables, graphs, maps or a photograph, and a series of multiple-choice questions is set:

QUESTION **16.4** Which word is most appropriate to describe the field layout in Fig. 16.1: mixed, linear, rectangular, random?

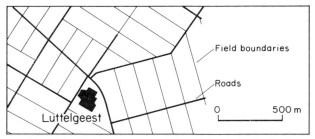

Fig. 16.1 *Field pattern on a new polder in the Netherlands*

Preparing for the examination 180

4 Some questions provide the setting for a problem. In this example, you are not asked to choose a reason as well as a location, but this is sometimes required:

QUESTION **16.5** Look at the map (Fig. 16.2). Which location is most likely to be chosen for a large, modern, planned shopping centre with plenty of car parking and accessible to a large number of people? Answer **A**, **B**, **C** or **D**.

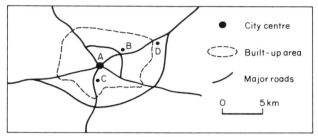

Fig. 16.2

5 Sometimes a multiple-choice question allows for more than one alternative to be chosen as the correct answer:

QUESTION **16.6** Study the map on page 180 (Fig. 16.3). What features are shown: (*a*) a meander, (*b*) a flood plain, (*c*) an ox-bow lake? Write down letter **A, B, C, D** or **E** according to which of the answers below is correct.

A only (*a*) is correct.
B only (*b*) is correct.
C only (*a*) and (*b*) are correct.
D only (*b*) and (*c*) are correct.
E (*a*), (*b*) and (*c*) are correct.

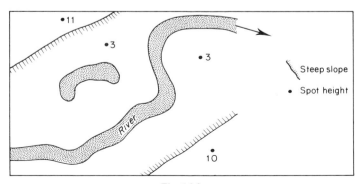

Fig. 16.3

181 *Preparing for the examination*

6 Another variation is to provide a set of information and ask you to choose the correct statement or generalisation from a list provided:

QUESTION **16.7** Study the graph showing the population structure of Egypt (Fig. 7.7(a), page 91).
Which of the following statements is correct?
A A large proportion of the population of Egypt is the older age groups.
B The greater proportion of Egypt's population is below 15 years of age.
C Egypt's population structure shows that most people are in the 20 to 50 age groups.
D The population pyramid shows an excess of females over males.

7 'Statement and reason' questions give statements which may or may not be correct, together with matching reasons which may or may not explain the statement. You have to select the correct combination.

QUESTION **16.8** Write down the letter **A, B, C** or **D** according to whether the pairs of statements and reasons below are correct or false:

	Statement		Reason
(a)	Egypt's population is growing rapidly	because	Birth rates have fallen quickly
(b)	Snow lasts longer on mountains	because	Temperatures decrease with altitude

	Statement	Reason
A	Correct	False
B	False	Correct
C	False	False
D	Correct	Correct

As you can see, the instructions for some multiple-choice questions need to be read very carefully.

Long questions
Longer questions take about 25 to 35 minutes each to answer, depending on your examination board. They come in two basic types:

1 Data-response questions, which provide information in various ways and ask detailed questions based on that information. The questions generally test your ability to apply your knowledge and understanding to the particular example given. However, some of the question parts may call for details of other examples or for you to give explanations of terms used.
2 Questions which are partly data response, but which also call for a fairly detailed answer using your own example or examples.

There are many examples of long questions throughout this book. Here are two more:

Preparing for the examination 182

QUESTION **16.9**
(a) Study the following newspaper extract: 'Another effect of June's heavy outburst has been noticeable chiefly on ploughlands, where flash floods have washed away the finer soil particles into drains and streams.'
  (i) What is meant by 'flash floods'?
  (ii) Describe and explain the process of soil erosion mentioned in the extract.
(b) For a region or area you have studied, describe the environmental problems created by changes in methods of farming or some other change.

QUESTION **16.10** The map and section in Fig. 16.4 show the typical patterns of land use of villages in the Dry Zone of Sri Lanka (the Dry Zone is the part of the country with a definite dry season and wet season).
(a) Describe the pattern of land use with reference to *intensity of land use*.
(b) Explain why the Old Field is usually cropped in both the wet and the dry seasons.
(c) Explain why extension of village farmland takes place along the valley.
(d) Describe and explain the pattern of land use of an area you have studied.

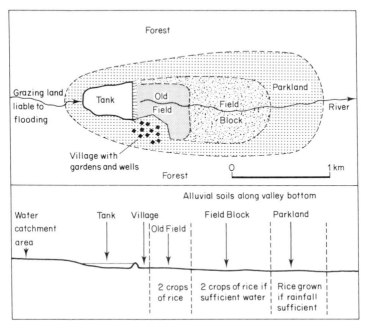

Fig. 16.4 *Land use of a village in the Dry Zone of Sri Lanka*

Answering the question

Many people lose marks by not paying attention to the *wording* of the question. For example, if a question asks you to *describe* the distribution of population, there is no point in giving reasons unless you are asked. Conversely, if you are asked to describe *and* explain, but you only give a description, you will lose half the marks for the question. In an examination there is a simple rule: *make sure you do what you are asked, but do not do more than you are asked.*

Throughout this book, there have been many examples of questions using various forms of wording. Here is a list of the common instructions. Make sure you know what is wanted from each:

Describe . . . Whatever you have to describe, keep to the point and set out your account in a logical order.
Explain . . . Make sure you keep your explanation to the point of the question.
Give reasons for . . . Make sure they are reasons and not a description.
Compare . . . What have they in common?
What are the similarities . . . ? What have they in common?
Contrast . . . How do they *differ*?
What are the differences between . . . ? Remember, *differences*.
Comment . . . Describe and give reasons for, and locate and name examples.
Write briefly . . . Describe and explain (and name examples if the question is not specific).
Suggest reasons for . . . Usually used where there is a range of possibilities; any reasonable or sensible points will be accepted.
Account for . . . The same as 'give reasons' or 'explain'.
Name . . . Usually an example you have to describe. Don't forget to name it.
Give the evidence for . . . State the facts to support a point, either from your knowledge or from data given in the question.
Locate . . . Say where a place or feature is (precisely if in map reading). May refer to your example.
Assess . . . Weigh up the importance of . . .

Here is a question which includes quite a number of these instructions:

QUESTION 16.11
(a) Describe and give reasons for the distribution of housing of different ages in a town you have studied. Name and locate your example.
(b) Compare and contrast the quality of the environment in two areas of the town you referred to in part (a). Suggest reasons for any differences you mention.

16

MAKING THE GRADE
16.3 Revision

1 Make a checklist for each topic in your syllabus. Write down each sub-section with the examples you studied. As you revise them, you can tick them off.
2 Revision means learning thoroughly. Don't just read through your work: make your own summaries of the main points of each topic to prove you know the work. Get someone to test you on the facts about your examples.
3 Practise answering examination questions so that you learn to judge how much time to spend on each part and still finish in the right time.
4 Practise drawing sketch maps, especially of your examples, so that you can reproduce them quickly in the examination. A good way to revise is to label your sketch maps with all the information about the example.
5 Keep referring to your atlas to improve your general knowledge of the world as a whole and individual regions specified in your syllabus.
6 Don't forget to carry on revising practical skills, especially map reading.
7 Check the way the examination paper is set out and be sure you do not miss revising a vital topic.

MAKING THE GRADE
16.4 In the examination
16

1 Note how many questions you have to answer and how long you should spend on each one.
2 Make sure you have everything you might need – pen, pencil, rubber, coloured pencils, ruler and possibly compass, protractor and set square.
3 If you have reading time before being allowed to start writing, make sure you use it well. Read the instructions carefully and then read the questions carefully. Questions which seem very difficult at first sight often turn out to be quite straightforward after some thought.
4 Decide which questions you are going to answer in the sections where you have to make a choice. Before starting a question, make sure you can do the whole question. It will be too late half way through.
5 Keep an eye on the time. Do not spend extra time on one question at the expense of the others. You are certain to lose more marks than you will gain by doing this.
6 Aim to finish with a few minutes to spare. This gives you time to check over your work and to make small corrections.
7 When answering a question, use your time sensibly. Do not spend 10 minutes out of 30 minutes on a part of the question worth only a few marks. If you have an answer paper with spaces for you to write in, you can use the spaces as a rough guide. Otherwise, use the number of marks for the question part (e.g. if part of the question is worth half the total marks, spend about half the time on it).

ACKNOWLEDGEMENTS

The author would like to thank the following for providing information for use in this book:
British Home Stores plc; Messrs R. and S. Cox; The Director, Brent Cross Shopping Centre; Mr Ian Forster; Kodak Ltd; Richard Marrat; Jaspal Singh; Martin Tudor.

The author and publishers are grateful to the following for permission to reproduce material in this book:

The *Ordnance Survey*, for four extracts (including the colour extract inside cover) reproduced from the Ordnance Survey 1:50000 map with the permission of the Controller of Her Majesty's Stationery Office, Crown copyright reserved; to the *Observer*, for one extract (p. 77); to *Penman & Partners Limited* for one extract (p. 136) from an advertisement for the Corby Industrial Development Centre.